THE

COTTON PLANTER'S MANUAL:

BEING A

COMPILATION OF FACTS

FROM THE BEST AUTHORITIES

ON THE

CULTURE OF COTTON;

ITS NATURAL HISTORY, CHEMICAL ANALYSIS,
TRADE, AND CONSUMPTION;

AND EMBRACING A

HISTORY OF COTTON AND THE COTTON GIN.

BY J. A. TURNER.

NEW YORK:
C. M. SAXTON AND COMPANY,
AGRICULTURAL BOOK PUBLISHERS,
No. 140 FULTON STREET.
1857.

EDWARD O. JENKINS,
Printer and Stereotyper,
26 Frankfort St.

PREFACE.

THIS book is a compilation. It makes no pretensions whatever to originality. All compilations must, from the very nature of things, be imperfect; therefore this book is imperfect. One of two plans I had to adopt, either to write an entirely original work, or compile one from the writings of others. Had I adopted the former plan, I might, it is true, have produced a more compact work, a more systematic treatise, in a more uniform style; but it is a question whether I would have made so valuable a volume as the one I present you. What might have been gained in the graces of composition, or the system of a well-digested treatise, might have been lost in my want of experience in all the departments I have presented you, to make a proper volume. It is quite easy to fill a given number of pages, but to make those pages useful and practical, is quite another thing. I thought it best, therefore, because it would be more useful to cotton planters, to compile the best authorities on the subject of which I treat.

The difficulties of selecting from such a mass of writings as I have had before me, and of so arranging the selections when made, as to form of them a compact volume, will be appreciated by every one—especially by those who have had experience in compiling; and yet I must be permitted to say that I think I have so far succeeded in the task I undertook, as to have given you all the most important knowledge which has been arrived at, with reference to the culture, consumption, and trade of one of the most important staples produced in the wide fields of agricultural labor. Not only is this an important book to the cotton planter, but to almost every class it will bear knowledge which it will be useful to have. The general reader will find in its pages many things with which he would be pleased to be acquainted.

This book is divided into eight chapters, as follows:

In a compilation such as this, it will not be surprising if the different chapters sometimes run into each other; that is, in some of the articles which are published under one head, there may be found some things which would more appropriately fall under another head. The locality of each article, however, had to be determined by the complexion of the major part of that article.

It will be seen that I have given the article of Professor McKay, on the cotton trade, from the year 1825 to 1850, both inclusive. Then follows a document, compiled in the State Department of the United States, which takes up the cotton trade where Professor McKay left it, and brings it down to the year 1855.

The articles which go to make up this volume, I have compiled from various sources. I am under particular obligations to the *Southern Cultivator*, *Soil of the South*, and *American Cotton Planter*—especially the first.

Errors have, in all probability, crept into this work. I invite communications from the pens of my brother planters, so that I may make a more complete treatise in a future edition of this book, or another volume.

I trust that the planting interest of the South will liberally patronize the publishers who have so liberally stepped forward to give them the first volume ever published on the culture of their all-important staple.

J. A. TURNER.

Turnwold, Putnam Co., Ga.,
Dec. 10, 1856.

CONTENTS.

CHAPTER I.

THE ORDINARY METHODS OF COTTON CULTURE.

CHAPTER II.

DR. N. B. CLOUD'S IMPROVED SYSTEM OF COTTON CULTURE.

CHAPTER III.

NATURAL HISTORY OF COTTON—ITS SPECIES AND VARIETIES. 

CHAPTER IV.

DISEASES AND INSECTS DESTRUCTIVE OF THE COTTON PLANT.

CHAPTER V.

ANALYSES OF THE COTTON PLANT, WITH SUGGESTIONS AS TO MANURES, ETC.

CHAPTER VI.

COTTON CONSUMPTION AND COTTON TRADE—COTTON TRADE FROM 1825 TO 1850. BY PROFESSOR McKAY, LATE OF THE UNIVERSITY OF GEORGIA.

CHAPTER VII.

LETTER FROM THE SECRETARY OF STATE,

CHAPTER VIII.

HISTORY OF COTTON AND THE COTTON GIN.

COTTON PLANTER'S MANUAL.

CHAPTER I.

SECTION I.—CHAMBERS' PREMIUM ESSAY.

THE following "Essay on the Treatment and Cultivation of Cotton," was read before the Southern Central Agricultural Association of Georgia, in 1852, by Col. James M. Chambers, then editor of the "*Soil of the South*," an agricultural journal, published in Columbus, Ga. It took the premium which was offered by the Association for the best Essay on the subject of which it treats. No more fitting article could be found with which to open this work. Col. Chambers is an intelligent man, and has always been an eminently practical and successful planter. Implicit confidence may be placed in his views.

The cotton plant is hard to be suited, in soil and in climate, and it rarely happens that such a combination of both is obtained, as to perfect the plant and mature the crop. The consequence is, that few spots are found, where these results are obtained with any degree of uniform success; but these do exist, to just such an extent as to demonstrate most conclusively that soils in proper localities are to be found, exactly suited to the successful culture of this delicate plant. With a knowledge of this fact, it becomes a matter of prime importance, to understand what these peculiarities of soil are, and where deficiencies exist, to search them out—and by artificial means, as far as it may be practicable, to correct or cure these defects of the soil in its natural state. We may not hope to remedy all the imperfections, yet it is the province of the cultivator to approximate as nearly as possible, and by preparation and culture, to endeavor to meet these peculiar wants of the plant. The first inquiry which presents itself is, to know what are the peculiarities of those soils which suit the growth and maturity of cotton. Experience is perhaps the safest and most reliable test, in the settlement of this question, and it is now pretty universally conceded, that our best cotton lands are those which are of deep and soft mold, a sort of medium between the sandy and spongy, and those soils which are hard and close—those which are penetrated by the warming rays of the sun, imbibing readily the stimulating gases of the atmosphere, and which allow the excess of rain water to settle so deep into the earth, as to lie at a harmless distance below the roots of the young plant. These are the properties of soil needful to the vigorous growth and early maturity of the cotton plant; and the knowledge of this fact is of great, and perhaps I might add, indispensable importance, to its successful cultivation. For though we may not find, and indeed it is very improbable that we should often find, all these essentials in the selection of a farm, yet by the aid of the plough, the hoe, and the spade, and the incorporation of foreign substances, we may remedy many defects, and supply many of the peculiar demands of this plant. These are all preliminaries to be arranged and understood, and from this point, we set out to discuss the question, as to the best methods of cultivating the cotton crop. It may already have been inferred, but I am not willing to leave it to inference, but make the assertion, that in my opinion, the best and most important part of the work, in cotton making, consists in a judicious and proper preparation of the soil for planting. It is difficult to say in all cases, and in the varied condition in which lands are found, and the diversity of soils, what the process of preparation should be; but we may lay down general principles for our government, and results to be obtained, and leave the planters to the selection of the best means at command for their accomplishment. All lands for cotton, ought, before the crop is planted, to be broken deep, close and soft, and this to be done long enough before planting, to allow the rains gently to settle them. It is the most common, and perhaps the best plan, to prepare all lands intended for cotton, in beds made by the turning-plough; and in flat and wet lands, sometimes an additional elevation ought to be given, by drawing up the beds with the hoe. I think, in this work, we have often followed too much the example of our neighbor, and have looked too little to reason, in the indiscriminate bedding and high elevation of all lands. I am the advocate of deep soft beds, made by very thorough and close ploughing, but cannot consent to the necessity or benefit, of elevating much, lands which are warm and dry, and which are not subject to inundations from excessive rains. For the convenience of culture, I would have the young cotton stand on a slight elevation, but when the condition of the land did not require it, I would not give it more.

The *distance* to be given is the next inquiry to be considered. This is a very important object, and one upon which we are very dependent for success, and yet it must be varied very much by circumstances, some of which are beyond our knowledge or control. The general principle may be stated, and then our best judgment must guide us in its application.

When the crop is at maturity, the branches of the stalks ought slightly to interlock every way. We cannot, therefore, do better in planting, than make an estimate of the probable average size to which the weed will grow, dependent, of course upon the vicissitudes of the seasons. It would, therefore, be vain to attempt to be more specific in directions, which must be varied always to suit the varied character of the soil. This whole question then, is to be settled upon the principle already stated. The planting should be in drills, chiefly because of the difficulty of obtaining good stands in hills; and I would add, for the information of those who may be without experience, that in the common medium lands of the country, these rows ought ordinarily to be about four feet apart, and the stalks in the drill should be thinned, so as to stand from fifteen to twenty inches from each other. The width of the rows and the distance in the drill, may be increased upon better lands, and in some cases of very thin lands, it may fall a little below the distances designated. I do not regard it a matter of indispensable importance, but should decidedly prefer that the rows should run in such direction as to give the plant the largest benefit of the sun from early morn to its setting. The cotton is decidedly a sun plant.

The Mode of Planting.—Here we have many plans, all setting up claims to some peculiar merit. With the preparation which I have indicated, it would hardly be necessary to stop to discuss the relative merits of these modes, or seek to do more for the accomplishment of our purpose, than to select some one, which we know to answer well. I therefore advise the use of some small and very narrow plough, for the opening furrow. This should be run in the centre of the bed, opening a straight furrow of uniform size and depth. In this the seed should be strewed by some careful hand, scattering them uniformly along the furrow, just thick enough to secure a good stand the whole length of the row. These I would cover with a board, made of some hard wood, an inch or an inch and a-half thick, about eight inches broad, and thirty inches long, beveled on the lower edge so as to make it sharp, slightly notched in the middle so as to straddle the row, with a hole bored in the centre one inch from the upper edge, and screwed on the foot of a common shovel or scooter plough stock. This wooden scraper and coverer, when drawn over the row, covers the seed nicely, leaving a slight elevation to prevent the settling of water, and dresses the whole surface of the bed neatly, for the space of fifteen inches on each side of the drill. Thus all clods or obstructions are removed, and a clean space is left wide enough for the passage of the plough in the first working between the young cotton and the rough land. This is an advantage of much importance with a crop so tender and small as cotton at this stage.

I have now conducted the operator, by a regular series, to the closing operation of planting the crop. And here I may be permitted to remark, that fine returns are sometimes obtained with much less preparation. These are results from the accidents of season, and not the due reward of well-directed culture; a prize drawn from the lottery, against which there are many blanks; a demonstration of the futility and uncertainty of all the best laid schemes of man. I pause before taking the next step. In this age of improvement, with scrapers and cultivators, and all the endless variety of labor-saving ploughs, and amid advocates for hard-culture and soft-culture, and high-ways and by-ways, for making the crop, "who shall resolve the doubt when all pretend to know?" and who shall decide, with such differences among doctors, who is right? and who can pretend to say what number of acres to a hand will constitute a crop with such varied modes of culture? I shall proceed upon the supposition, that a plentiful supply of provisions are to be made on the farm, and then set down as a good cotton crop, ten acres to the hand;

under favorable circumstances, a little more may be cultivated, and on some lands less. Upon this basis, I proceed. As soon as the young cotton is up to a good stand, and the third and fourth leaves begin to appear, the operation may commence. In lands which are smooth and soft, I incline to the opinion, that the hoes should precede the ploughs, chopping into bunches, passing very rapidly on, and let a careful ploughman follow, on each side of the drill, throwing a little light dirt into the spaces made by the hoe, and a little also about the roots of the cotton, *covering and leaving covered*, all small grass which may have sprung up. This is, indeed, the merit claimed for the operation, that after the hoes have passed, the ploughs come on and effectually cover and destroy the coat of young grass then up. This is known to practical planters, to be the crop of grass which escapes the hoe, and does mischief to the cotton. But when the land is so rough as to endanger the covering of the cotton with the plough, the operation must be reversed, and the hoes follow the ploughs. As all that is now proposed to be done is, a very rapid superficial working, reducing the crop to bunches, soon to pass over and return again, for a more careful operation. This should be done as soon as possible, as will be indicated by the necessities of the case. The grass and the weeds must be kept down, and the stand of cotton reduced. At this first working, unless in lands already very soft, I should advise the siding to be close, and to be done with some plough which would break and loosen the earth deep about the roots of the young plant. Others may theorize as they choose, but with a plant sending out a tap root, upon which it so much relies, and striking so deep into the earth, as that of cotton, I shall insist upon its accommodation, by providing a soft, deep, mellow bed, into which these roots may easily penetrate. In the second working, the ploughs should in all cases go before the hoes, and in all lands at all tenacious or hard, let the work be deep and close again, and the middles of the row also be well broken up at this time. Now the hoes have an important and delicate duty to perform. The cotton is to be reduced nearly to a stand, though it is now rather early to be fully reduced. It is perhaps best to leave two stalks where one is intended to grow. The young stalk is very tender, and easily injured, by bruises and skins from rough and careless work, and it is much better to aid a little sometimes with the hand in thinning, than to spoil a good stand, by bruises from the hoe. The cut-worm and the louse are charged with many sins, which ought to be put down to the account of careless working, at this critical stage of the crop. The distance to be given I have before stated, and in the first operation of bunching, this ought to be looked to, and the spaces regulated accordingly. At this second passing over, the hoes must return a little soft dirt to the foot of the stalk, leaving it clean and supported. If this work is well done, the weed will grow on, without any necessity for further attention for some twenty days or three weeks, when the plough should return again. At this time, some plough should be used next the cotton, which will tumble the soft earth about the root, covering the small young grass, which may have sprung up since the last working, but the ploughing should be less close, and shallower, than at the former working. The hoes have much to do in the culture of this crop, and must be prepared to devote pretty much all their time to it, constantly passing over, and perfecting that which cannot be done with the ploughs, by thinning out surplus stalks, cleaning away remaining bunches of grass, stirring about the roots of the plant, and if need be, adding a little earth to them. It is difficult, in a treatise of this sort, to say how often, and in what manner, this crop shall always be worked, when the character of the seasons, and the difference in the land, must have necessarily so much to do in settling this question. The general rule must be, to keep the earth loose and well stirred; the early workings to be deep and close; and as the crop comes on and the fruit begins to appear, let these workings be less close, and shallower, keeping the soil soft and clean. It is of great importance to work this crop late, and it should not cease until the branches lock or the cotton begins to open. I do not consider that it is necessary to pile the earth in large quantities about the roots of the cotton, but think the tendency of all the workings should be, to increase the quantity.

The selection of seed is an interest not to be disregarded. We have been humbugged a great deal by dealers and speculators in this article, yet we would greatly err to conclude that no improvement could be made. We should, however, save ourselves from this sort of imposition, and improve our own seed, by going into the field and picking each year, from some of the best formed and best bearing stalks, and thus keep up the improvement. Great benefits may often be derived, by changes of seed in the same neighborhood, from differences of soil, and occasional changes from a distant and different climate may be made to great advantage.

The picking of cotton should commence just as soon as the hands can be at all profitably employed—say as soon as forty or fifty pounds to the hand can be gathered. It is of great importance, not only to the success of the work, but to the complexion and character of the staple, to keep well up, with this work, so that, as far as possible, it may be saved without exposure to rain. The embarrassments to picking when once behind, and a storm or heavy rain shall intervene, mingling it with the leaf, and tangling in the burr, are just as great, as to get behind it in the cultivation of the crop, when much additional labor will be required to accomplish the same object.

In the early pickings, when the seeds are green, some sunning is indispensably necessary; but after more maturity and dryness, very little will be required. This must be determined very much by circumstances; but dew or rain-water should always be removed, by drying upon the scaffold, before the cotton is bulked in the house. With proper care and attention, great improvement may be given to the complexion of the staple by a little heating in the bulk, extracting the oil from the seed, and imparting a slight cream to the color. This process, however, must be conducted with great caution and care, lest the heating proceed too far, and injury be done. It is easily checked, by stirring and exposure to the air. It is an advantage to all cotton to lie in the bulk before ginning, and we doubtless often lose much of this benefit for want of sufficient house-room. Indeed, I think it a very common error, in our plantation arrangements, not to build houses for this special object. The cotton, when ginned, ought to be so dry, that the seed will crack when pressed between the teeth. It is often ginned wetter, but just as often, the cotton samples blue. A gin should be used which will neither cut nor nap the cotton, but send out the fibre straight and smooth, so that when the samples are drawn, they will have the appearance of having been carded. This is greatly promoted by the largely increased number of brushes now added by the best manufacturers.

The packing should be in square bales; and, without reference to freight, or any of these mere incidental influences, I think the weight of the bale should be fixed at about four hundred, or four hundred and twenty-five pounds; to be in two breadths of wide bagging, pressed until the side seams are well closed, or a little lapped, and then secured with six good ropes, the heads neatly sewed in, so that when complete and turned out of the press, no cotton should be seen exposed. These packages should be nearly square, for the greater beauty of the bales, but still more, for the greater convenience with which they may be handled and shipped, saving the necessity for tearing the bags, and giving a better guarantee that they will reach a distant market in good order.

The crop is now made and ready for market, and as I have gone through with the labor of making, I hope I may be pardoned for manifesting a little interest as to its disposal, and therefore venture to offer a little advice on that subject. Create no liens on this crop, or necessity for selling. Never spend the money which it is to produce, until it is sold. You are then free to choose your own market, and time of selling; and as cotton is a controlling article, it will generally regulate the value of all property to be purchased, *except the redemption of an outstanding promise.*

I might have said something about the topping of cotton, but all I could have done, would have been to put it down as a contingent operation, and doubtful in its effects upon the crop, I might also have descanted largely in the enumeration and description of insects and diseases peculiar to cotton, suggested some remedy, and swelled my essay, by a flourish in the dark, upon topics about which little is known; but I have felt that it would be most in accordance with my plan, and certainly most with my feelings, to candidly confess my inability, and include these all under the head of Providential contingencies, to which this crop is liable, and against which we may war and contend, but which will, after all, prove an overmatch for the energy, skill, or wisdom of man."

SECTION II.—GOVERNOR HAMMOND'S REPORT.

From the South-Western Farmer.

WE give at the conclusion of this notice, the Report of a Committee of which our friend, J. H. Hammond, was chairman. We congratulate him on the knowledge of farming that he displays. We see how readily the educated and intelligent can learn the business of farming. But a short time since, our old school-mate was up to his eyes in politics—he now retires to the field—there to live a quiet, peaceable life. We rejoice at it, and can but repeat the remark we made to him, before he was made Governor—"You are wrong—you have no business in that sphere—seek your ease and peace—it suits you better, and will give you satisfaction." His answer was then, as his works answer now: "I will do so as early as the force of circumstances will permit"—or to this purport were both.

We again congratulate him, and also our country, in the success of our friend—we also press on all agriculturists any article from the pen of Hammond; he will, we feel assured, give all matters that he writes on, his minute and particular attention. We have known him from both of our boyhoods, and know him to be talented and observing, and more than all, when he does apply himself, it is an application deserving and insuring success.

As we are his senior in planting the cotton, especially in personal attention to it, we beg to give him a hint or two. We may err in our notions; and why we say so, is, that we differ materially from so large a number of farmers. We think that very early planting is disadvantageous; and to define early planting, we think the last week in March is early enough at any time, even for this year, when, it will be borne in mind, the fruit trees were quite green at that time; to plant as early as the 15th or 20th of March is "very early." We generally judge it to be time to plant corn when the "leaf of the oak is as large as the squirrel's ear;" many of our planters have planted cotton as early. We think cotton planted from the 1st to the 10th of April is early enough for old land, and have known by several crops that the later planting, say 10th, was considerably better than the early—we know this not only by our own weights and measures, but also by others.

We would make an exception to early planting. New ground and rich fresh land, has such a tendency to make weed, that it is necessary to plant as early as a stand can be had—so all think; we would not object, but think that judicious culture would make a different result. We would act precisely as with the tree that produced wood instead of fruit—amputate the roots. We think that if the land has been broken up very early, and left to be consolidated by rains, then planted about the 5th to the 10th of April, thinned out as early as it was up, cultivate deep and late, that the stalks would set the fruit and ripen in time. Do we not thus with fruit trees, Irish potatoes, and sweet potatoes—the latter two, by either cutting off tops, or feeding with calves? And why not a similar practice with the cotton plant?

The cotton plant is a very tender plant if treated as it was some ten or fifteen years since—some three to six bushels of seed sown per acre, and not thinned out until the third leaf had appeared. It has been raised in a hot-bed, and no wonder it is tender; but if sown thinly, and then thinned out to single stalks, we think it a hardy plant. There is no use in trying the hardihood of the plant. It is unlike corn—it has a tap root, grows in dry weather, and unless the land has not been properly prepared, or remarkably dry, it will improve by hot or dry weather—but corn having superficial roots, should be planted early as possible, that it may ripen before drought sets in. If cotton will make 1000 lbs. per acre, when planted late in May, there can be no fears to plant 10th April. The farmer can place his land in excellent order—have his corn cleaned handsomely, and when cotton is up, he can rush it to the utmost. We request our friend H. to plant one acre of cotton, even now, after his seeing this, on a piece of well-ploughed land, in the same field that he has now even scraped over; just open out furrows where the cotton now stands, which will destroy the cotton that has been scraped. Our impression is, that the difference will be very slight, and if adopted generally would give considerable time to manure, plough, and improve, instead of giving cotton the additional working necessary. The land that we have known planted late, would not, in average seasons, make anything like one-third more, by early planting, and if the extra labor was applied to improving it, we doubt, if it would yield as much. Understand, we do not advocate either late or early planting—that is, after the 25th of April or before 1st—and only wish to show there is not so vast a difference between planting 1st of April and 1st of May. What would be the gain to any farm by the extra month's work?

Report of the Committee of the Barnwell Agricultural Society, on the Culture of Cotton.

The ground cannot be too well prepared for Cotton.—If it had rested one year, it should be broken flush, as early in the previous fall as possible, and spaded just before planting. If it has rested two years, or been planted the preceding year, let it be listed as early as it can be done, and two furrows thrown upon the list. Immediately upon planting, let two more furrows be thrown up, and balk broken out completely. The common method of running three furrows, and planting on it, throws the winter's portion of the crop-work upon the laborer, during crop time, and is inexcusable, unless heavy clearings are absolutely required. The reason for not listing after one year's rest, is, that the vegetable matter will be too abundant and too coarse to form a substratum to receive the tap root.

Cotton should be planted early.—It may increase the difficulty of getting a stand, and give the plant, for a long time, a puny appearance, but every stalk of cotton planted in March, or first week in April, that survives, may be readily distinguished, in any field that has been replanted later. It bears more, and earlier, and stands all the vicissitudes of June, July and August better. There are several methods of planting. Your Committee recommends planting in spots, regularly measured by the dibble. It is somewhat tedious, though less so than is generally supposed, and certainly does not take as much time as both to drill and chop out; nor is time so valuable at that period, as when the latter operation is required, while a better and more regular stand may be secured. There is no land, or but little, in our district, in which cotton rows should be over three feet apart, or the cotton further than fourteen inches in the drill, one plant in a place. To make a large crop, there must be an abundant supply of stalks. When the weather is too wet to plant, time may be often saved by dropping the seed, but not covering until the ground is dryer. If, however, it cannot be covered in three or four days, it is time lost, for it must be replanted. Always cover lightly, under any circumstances. And always plant on something of a bed, in any land. It keeps cotton dryer, and affords more air when it is young. It enables you to get at it in working. By increasing the surface, it absorbs more moisture, if it is too dry; and gives out more if it is too wet, and in both cases, gives you the advantage of a vertical sun on the tap root, which hastens the maturity of the bolls—a vast desideratum in our climate. On this account, the bed can hardly be drawn too high at the last hoeing, in any season.

In cultivating Cotton, whether with the plough or hoe, the chief object is to keep down the grass, which is its greatest antagonist, bringing all, or almost all other evils in its train. It is not so essential, in the opinion of your Committee, to keep the ground stirred, as is generally supposed, and by no means requisite to stir it deep ; at all events, not to our light soil. If it be well prepared, deep ploughing is not only unnecessary for any of our crops, but often highly injurious to them, while it rapidly exhausts the land, by turning it up fresh, under a burning sun. Much unnecessary pains is usually taken, and time lost, to work the plant in a particular way, under the supposition that it is a peculiarly delicate one. If it survives its infancy, few plants are hardier. It is often found to reach maturity in the alleys, where the mules walk with the ploughs following, and the laborer tramps backward and forward. Sometimes it will bear fruit in turnrows used frequently for wagons, while it really seems to derive benefit from being bitten down almost to the ground by the animals; it will bear almost any usage better than it will that mortal enemy—grass.

The most critical operation in working cotton, *is thinning.* It should be done with great care, and if early, with the hand. In a dry year, it cannot be done too early after the plant is up. In a wet one, it may be profitably delayed, until it has begun to form, or later even. On the experience, observation and judgment of the planter, in this matter, everything depends, as each year brings its own rules with it. Where circumstances are favorable, early thinning is, of course, the best. Some planters always top their cotton. Others never do. Your Committee are of opinion, that it seldom or never does harm to do so. But whether it is worth the trouble, is a doubtful question. Those who have no clearing, or other important employ for their hands, would lose nothing by devoting three or four days to this operation early in August. Those pressed for time might gain by omitting it.

Too much pains cannot be taken in preparing Cotton for market, for they are well remunerated by the additional price. The first thing to be attended to, is to have it gathered free of trash. With a little care, wonders can be effected in this way; and hands with a short training, will pick almost, if not quite, as much without trash as with it. It should never be gathered when wet. And here it may not be out of place to remark, that one of the very best sanitary rules of a plantation is, never to send out your hands to pick until the dew has nearly or quite disappeared. It saves time in the long run, as well as health and life. Cotton should never be ginned, until the seeds are so dry as to crack between the teeth. If damp, it is preferable to dry it in the shade, as the sun extracts the oil, and injures the staple. If, by accident, however, it gets wet, there is no alternative but to put it on the scaffold. It is of great importance to sort the cotton carefully into several qualities, in ginning and packing, for by mixing all qualities together, the average of the price is certainly lowered. A few old hands, or very young ones, breeding women, sucklers, and invalids, will earn excellent wages in a ginhouse, at this occupation. Neat packing is of no small importance, in the sale of cotton, and no little taste may be displayed in making the packages. The advantage of square bags is universally known, and the Committee are astonished that any other should ever be made now.

Every kind of manure is valuable for Cotton.—Every kind of compost, green crops turned in, cotton seed, and even naked leaves listed and left to rot, improves this crop. When planted on cotton seed, and sometimes on strong stable manure, it is more difficult to retain a stand, owing probably to the over stimulus of these strong manures. So, on leaves, unless well rotted, the cotton will long continue to die, in consequence of the leaves decaying away, and exposing the root too much to sun and rain. These difficulties may be avoided, by a little pains, and by no means justify the opinion entertained by some, that cotton should never be planted on freshly manured land. The only question is the cost of the manure. A great deal may be made on every plantation, without much trouble or expense, by keeping the stables and stable-yard, hog and cow pens, well supplied with leaves and straw. And also from pens of corn-cobs; sweepings from negro and fowl-house yards, and rank weeds that spring up about them, collected together, and left to rot. Whenever the business is carried further, and a regular force is detached to make manure at all seasons, and entirely left out from the crop, it becomes the owner to enter into a close calculation of the cost and profits. In many agricultural operations, such a course, the experience of all countries has proved to be profitable, but these operations partake rather more of the farming and gardening, than planting character; and whether the same method will do for the extensive planting of short staple cotton, remains, in the opinion of your Committee, yet to be tested. If anything like an average of past prices can be maintained, it is certain that more can be made by planting largely than by making manure as a crop. If, however, prices continue to fall, and the growing of cotton be confined to a few rich spots—those susceptible of high manuring—then our whole system must be changed, our crops must be curtailed, and staple-labor losing its past value, the comparative profit of a cotton and manure crop, will preponderate in favor of the latter. As a substitute for manuring on a large scale, resting and rotation of crops is resorted to. In our right level land, the practice of resting cannot be too highly recommended, and, by a judicious course, such as resting two and planting two, or at most three years, our lands may not only be kept up for ever, but absolutely improved. From rotation of crops, but little is gained for cotton. After small grain, whether from the exhausting nature of that crop, on light lands, or because the stubble keeps the ground always rough and porous, cotton will not do well. After corn, it is difficult to tend, as from our usual manner of cultivating corn, grass is always left in full possession of the field. It does best after cotton, or after a year's rest. Rest is the grand restorer, and the rotation chiefly required in the cultivation of cotton.

J. H. HAMMOND, *Chairman.*

SECTION III.—"COLO" TO HON. J. C. CALHOUN, ON COTTON CULTURE.

From the Laurensville Herald.

HONORED SIR:—Will you permit me, through the columns of the *Herald*, to reply to your very acceptable letter.

The subject I desire to press home upon every planter is the improvement of seed by a close and rigid selection from the field, as also the duty of drying before put into bulk, so as to prevent the heating of seed.

Every planter should do it to some extent, and in addition thereto procure an occasional fresh article from the favored region of the cotton plant.

I commence my cotton planting operations by breaking down with clubs the cotton stalks of the past year: if they be large, the limbs are threshed down first so as to break up, then the stalk broken off as near the earth as possible. Of course this is done when cotton succeeds cotton. I then run off my rows, at such distances as the fertility and age of land as well as the variety of seed demand. The fresher the land, and richer it is, the greater the distance; the Mexican seed requiring more distance than the cotton I have seen, which is called in a part of Mississippi, the Hogan seed—a few I have received as a present—and these still more than the Sugar Loaf, another variety from Mississippi, which in some localities in the Gulf States has proved very productive. I have not had occasion to give a greater distance than five and a-half feet, and am inclined to think—though you claim to be at the northern extremity of the cotton region—that upon rich and fresh land the cotton stalk may be as large or larger than some eighty or ninety miles south, on similar lands.

I make it a point to plough out all land as deep as I can, and without any ridge being left under the ploughed land. My rows are always laid off by stakes, with a shovel-plough, and then two furrows turned to it, one from each side, with an efficient turned plough; this is performed as early in March as I can, endeavoring to postpone my spring ploughing until after the heavy rains. Understand, I have a clay subsoil, with silicious matter so fine, that no grit is perceived by rubbing with the fingers.

Using due diligence in my early ploughing, and planting of corn, I am enabled to have all cotton land with three furrows thrown up, before time to plant cotton. When the time has arrived—which time should not be before the seed will vegetate, and plant grow off—I do not like to plant as early as many do—I then press forward my ploughing and planting thus: enough ploughs go ahead to ridge up entirely the balance of unbroken earth; harrows follow, openers, droppers, and last coverers. I never wish to sow more than one bushel of seed, and prefer to cover with a board or block so as to cover shallow, to leave ridge smooth, and to compress earth to seed. Upon level land, I require a set of hands to plant ten acres per day, length of rows averaging four hundred and forty yards; a set of hands is one harrower, one opener one to sow seed, and one to cover. Now, esteemed sir, we have planted say one-half the crop.

If all the land had not been ploughed with three furrows prior to this, I then turn about and prepare the residue of land, and if corn can be pressed forward, I work all or part—with the view of having ten days between first and last planting. Then return to planting the residue of cotton. We have now planted the crop.

Ploughing and Planting.—I am very particular in requiring rows to be laid off straight, bedded up so; and furrows opened for dropping, equally so, because the ploughman in all succeeding labor is able to plough nearer the plant, thus lightening hoe labor. An expert ploughman, with a sharp turning-plough, by letting the share run *level with the ridge*, handles inclined, of course, can scrape so near the plant, that a hoe hand can scrape and thin out nearly twice as much.

Many, in breaking up land for cotton, leave unbroken earth; some call it "cut and cover"—that is, cover unbroken earth with a harrow—and they insist that the plant bears better than when the land is broken up—the plant grows too luxuriantly. This may possibly be the case upon the rich lands where your plantation in Alabama is, but certainly not in *our State*, and where you live. It is a slovenly culture, to say the least of it. But how can the tender spongioles of the root pass through stiff land in dry weather, and how can the plant be sustained when only half the land is cultivated?

The deeper land is ploughed when the subsoil is not sandy, or gravelly, if properly drained, the more room for roots to search for their food, and the greater deposit of dew therein, the longer to get hot, and the earlier to cool, as well as holding more moisture, less liability to wash from an ordinary rain, and the sooner the drying of the surface.

I place two furrows on the one laid off early, that the earth may consolidate—cotton seed vegetating more certain, and grows off more rapidly. I put off breaking out the residue as long as I can, so that the surface may be clean when planted, and thus grass and cotton have an equal start. I use the harrow to remove all trash, clods, &c., as also to level ridge.

I prefer a ridge, with the view of having dry, warm soil for the seed, as cotton grows off earlier, and is sooner out of the way of droughts, as also that I can scrape down with the plough, and cover young grass thinly in the middle.

Early planting gives "sore shin" and lice; or rather the plant has so little vitality, that its natural enemies soon "take away even that which it hath."

I always strive to keep seed perfectly sound, thereby adding to the vitality of the plant. I have noticed, some years the stand to be worse than other years, and some men always to have had the luck of bad stands. This was owing, *I think*, to damp weather, or wet spells injuring the cotton so as to injure the vital powers of the seed.

I plant seed sparsely, because the plant becomes hardy at once, and then stands almost, if not quite, as much cold as corn.

I regard a crop when planted in first-rate order, as nearly

half made, so much regard I place upon thorough tilth and thorough preparation.

With profound respect, I am, honored sir, yours,

COLO.

SECTION IV.—REMARKS ON THE CULTIVATION OF COTTON.

From the Southern Cultivator.

1. *Preparation of Land.*

In writing out the detailed plan I pursue in the cultivation of cotton, I must begin I suppose on the 1st of January, so as to carry your readers regularly through. I will endeavor not to be tedious, yet I cannot possibly be minute, without at least being tiresome to somebody—and there is always somebody, who already knows everything.

For ten years past, I have thrashed down all cotton stalks, cut down all corn stalks, and turned them under as well as possible with a turning-plough. When planting cotton after corn, I strive to break up the land with two-horse ploughs—what I term flushing, that is, breaking up in thirty to fifty-feet beds. Last year I broke up every acre of land I planted, with two-horse ploughs, whether planted in cotton, corn, oats, or potatoes.

If my land has been in cotton, I generally open out water furrows, deep, with a shovel-plough, to this I throw two furrows, one on each side, with one or two-horse turning-ploughs. Thus the land remains until a day or two before I wish to plant, when I have the balk broken out, thus having fresh earth to plant upon and yet firm earth for the seed to be planted in. There will be a narrow ridge of earth, not covered by the fresh earth, but I invariably run an iron-tooth harrow along the ridge, so as to break clods, and rake off pieces of stalk and to leave the ridge fresh; if once running of the harrow will not do, I run it twice.

The opener then follows and opens out a furrow, say one half inch is deep enough, and narrow; if this furrow could be as straight as a bee line, and half an inch wide I would esteem it better, if upon level land. The seed is scattered *thinly* and regularly, then covered with a board or block; I would prefer a roller. As to distance, this depends upon quality, age, and locality of land. Rich and fresh land requiring greater distance, and I am inclined to think that the same quality of land north of say 33° will tend more to longer joints, than does cotton about 31° to 33° and particularly western lands; these lands tending to short joints, and greater yield to height of cotton. I do not plant any land that requires rows to be over five and a-half feet, even to grow fifteen to twenty hundred-weight of cotton per acre. There is sometimes, I am sure, much loss by too sparse planting. I desire to have the plants meet in the rows by the 1st of August, and should it after this date lap in the row, the crop will not be materially injured. I find the new varieties, as Sugar Loaf and Cluster, to require less distance both ways than does the Mexican. When I planted my crop with Mexican—Petit Gulf—I gave five to five and a-half feet by two to three feet on my best land. For four years I have grown Sugar Loaf, and plant four and a-half feet by eighteen to twenty-four inches, preferring about eighteen inches. Upon second quality of land I reduce distance to four feet or less, by eighteen inches. Upon this department of planting (the preparation) I use more lime and labor than is usual, being careful to break up deep, throw out into beds all the land, leaving no unploughed ridges; the ridges I endeavor to pulverize well, and do not run ploughs unless land will pulverize, thinking ploughing may be done too early and land injured by being ploughed wet. My object in ploughing, say three furrows, early, is to permit the foundation of ridges to settle somewhat, as seed germinate freer, and grow off better than upon light earth. I break out the residue as late as planting time, so that the plant will start before or with the grass and weeds. I prefer never more than a bushel of seed, per acre because solitary stalks are not injured by cold weather when scraped out, as when grown in a hot-bed.

I have been asked, how I plant seed when I buy. I reply, I wet the seed thoroughly with salt and water, and sometimes use brine made by steeping stable manure in salt and water for ten days before wanted, until fermentation has ensued. The seed are then dried off with ashes, or lime, or plaster—I prefer the two latter, as the seeds are white, and the master can see that care in dropping is practised by hands. These seeds are dropped at the required distance and are covered with the foot, by brushing a little earth upon the seeds and pressing them into the earth with the foot. I would prefer a seed-planter, but could not make the one I tried, drop regular. Five to ten seeds in a place is ample. I have dropped only one, and two, and three ; when I did this myself, I failed not in a stand.

With a good ridge, clean of clods and litter, a hand can scrape more ; the labor of planting carefully, and time seemingly lost in this, as well as of dropping seed, is fully regained in the scraping. I have cultivated for ten years, nine to ten acres of cotton, and eight to nine of corn, besides potatoes, oats, &c. This could not have been done, but by doing all work well. Time is saved by good ploughing and neat planting.

2. *Preparation of Land and Planting.*

Last night, I gave you the preparation and planting of the cotton crop; yet I could not, in the length of one article, give more than a rapid survey. I prefer short articles, and yet it is best to be particular, even minute—though there is even here an objection—for a writer should leave something for his readers to think of. When I plant oats land, land that was the year previous to rest, or corn land, I invariably break up into large beds, size according to width of rows to be planted, so as to throw water-furrow of the flushing as a water-furrow of the row. When four feet rows, I run off land thirty-two feet, and keep furrows as straight as possible, on level land. I then lay off rows, always with a shovel-plough, and then two furrows as before. Sometimes I open out water-furrow of old rows, as deep as two mules can draw a shovel-plough; bed up to this entire, then open out a new water-furrow deep, and reverse two furrows with a one-horse plough. I am satisfied that there is no land I plant but what is materially benefited by breaking up with a two-horse plough, then bed up with a one-horse plough—thus all trash, grass, seed, &c., is well buried below the one-horse plough furrow. I use a piece of wood two to three feet long, running level on the land, the front end shod with iron, for the purpose of opening out furrows for planting seed. My object is to make a clean, straight furrow, and impact the loose earth. This stick of wood is rounded below, and fastened to a shovel-plough stock. The straighter the row on level land, or the more regular on rolling land, if circling be practised, the closer can the scraper be run—thus giving less labor to hoe hands. And if cotton seed be scattered very regular, so as to give a stand, no stalks touching, the hoe hand can thin out faster, and thus save time.

If I were able to plant my cotton crop with the neatness and order with which Col. Wade Hampton plants his crop, I believe I could cultivate an acre or two more per hand. Being in company with him in 1847, on a steamboat, we discussed the subject of planting for hours, and he assured me that all his furrows were opened out for planting with the corner of the hoe, narrow and straight. If I could drop seed in a furrow only an inch wide and quite straight, I think I could manage two acres of scraping per day to each full hand. I regard planting a crop, if done in the best manner, more in the light of half cultivated, than many would believe. I have scraped three acres in a day. I can dirt easily four acres per horse; and can, with the solid sweep, break out four to nine acres per horse, owing to whether rows be four or five feet wide—thus, besides the earthing furrow, it requires one or two to sweep out the middle. But land has to be put in good order, and seed planted in order. This matter has called for many a line from my pen in the different papers I have written for, and I must be pardoned for thus dwelling so long. It is really no interest of mine whether planters cultivate well or ill; whether they can cultivate a fair crop easily, or not, I cannot be benefited. Yet, as a citizen of this beautiful world—as a sojourner in this southern clime—I feel an abiding interest in the welfare of my fellows. Therefore, I say, if planters will devote more care and attention in tilling their lands, and in putting in their crops in a good manner, they will be able to make more, and yet spare their servants and their beasts much labor in the cultivation.

Look at the garden. Take one bed and trench it—spade up two spades deep, reversing the soil even, what will be the result? But suppose the first spit be laid one side, then the second spit well and finely dug up, the first returned reversed, or thoroughly mixed—will not that bed be more or less moist all the year? And if there is a chance for water to pass off, will it not be fit to work after a rain, sooner than any part of the garden? And must it not, of necessity, produce better?

I admit a planter cannot plant so great a crop, but he will need much less to make an equal crop.

The misfortune is, the body of the cotton planters want a large crop, and will not be at the expense of team and tools. Would they not ridicule the carpenter who, instead of getting tools to tongue and groove his flooring, would attempt to rabit each side of plank, or to dig grooves, and then dig for a tongue with a chisel? And yet, though not quite so absurd, planters act. What difference in cost, in twenty years, if a planter buys six shovels, six one-horse turning-ploughs, three two-horse turning-ploughs, six scrapers, six harrows, or to buy all turn-ploughs? These same ploughs will last by changing—those not used to be taken care of—as long as the same number of one kind, and for all work. Think ye, and judge ye.

3. *Cultivation of the Crop.*

Mr. Editor:—I have seen cotton cultivated these thirty or more years; I have read pretty much all that has been written upon the subject since 1819, and I have tried many experiments. So far, I know no better way to proceed in the culture than the plan I now pursue, and is pursued generally in this section. Who deserves the credit for the plan, I know not; nor do I know from whence it came. To cultivate a full crop, we must rely on the plough; if we use ploughs adapted to the work, I can see no objection. And, as to scraping cotton, the best planters of this part of Mississippi not only do it, but they are falling into the plan of even scraping corn.

But to our work. I never plant all cotton land at once; I prefer much to plant, say one-half about the 1st of April, and in ten days the residue.

So soon as I see enough cotton up to make a stand, I begin to scrape, by starting a scraper, which, if in good order and beds well made, will shave the bed to within one inch of the plant, and cover all the grass in the middle of a four-foot row. And here will be seen the great gain in throwing up a ridge and planting seed straight on a clean bed. Hoe hands follow, and scrape the entire remaining surface of the ridge, leaving none of the surface to grow grass, chopping through the row so as to leave either a stand at the right distance, say two or three, or four stalks, or else leaving a double stand. By beginning early, the surface is cleaner, and it is much less difficult to clean around the plant, than when grass and weeds have started. And here let me say, I used the best steel-bladed hoes sold by A. B. Allen, in New York, and by S. Franklin, in New Orleans. I do not like the largest size. These hoes can be ground on a grind-stone every two or three days, and a flat file used daily will keep them quite sharp.

My hoes having made a start, the earth being dry enough to crumble after a plough, I start bull-tongue ploughs immediately after, and dirt the cotton, endeavoring to finish in a day or so after the hoes; thus, I have my cotton clean and earthed; it is protected from grass, and the light earth protects from the cold nights in May. Besides this, the deep narrow furrows made by the bull-tongue plough serve to drain the narrow ridge left. This certainly gives a warmer bed for cotton, and in throwing light earth, grass vegetates much slower.

Sometimes I scrape the second time, though the plough not preceding; when I do not, and the earth is comparatively clean, I start a small shovel-plough again to dirt, and the hoes follow, to clean and level the earth on the ridge, making it a point to clean the row perfectly. Never cover grass unless it is very small.

After this, I use sweeps, cultivators, shovels, harrows, &c. —very seldom a turning-plough—and at every working with plough, throw a little earth; every working of hoe, scrape a little off, so as to keep my bed about the same height. Cotton requiring a dry bed, not a high ridge; I therefore never draw up earth with hoes.

After the first earthing, the main principle is to keep clean and stir the earth every fourteen to twenty days; and continue this even to the picking, if on good, light, moist land. Better to break a few limbs of the plant, than to stop the ploughs too early.

I am opposed to waiting for the plant to have the third or fifth leaf before scraping; too much time is lost, grass gets some strength, and it is more tedious to clean the crop; besides, the plant is checked in growth, and almost invariably turns yellow after scraping. I also oppose scraping, if left two weeks before earthing the plant. I regard scraping as essential to the cleaning a crop of eight or ten acres per hand; but the plant should receive earth as soon after as possible. I have scraped and earthed with the hoe as I scraped, but this is again too slow. Scrape or clean off grass and weeds with the hoe, and dirt with the plough, is the principle. The object is to keep the young plant thrifty, that it may stand up against the louse. I would not speak as *ex cathedra*, nor in that tone; yet I give my opinions as is natural to me, using no *parlez vous* phrases.

As regards *lice* and *sore shin*, the first is a never-failing attendant of the cotton plant; but when the plant is healthy, the lice do not check the growth, nor do the things breed so fast. They are worse upon early planted or badly cultivated land, because the cold checks the growth, and if cultivated in mud and water, a similar result. *Sore shin*, I think, is caused by injury from the hoe, and cultivation when land is wet. Some seasons, both of these are worse than others. An early, warm season, with sound seed, we are but little annoyed; thus showing the fault is our own. I have never been annoyed but one year, and then I had a very smart overseer, who *got ready* to plant before he *was* ready.

4. *Cultivation of the Crop—continued.*

Mr. Editor:—There are many requisites to making a good crop, and the most of them are within the reach of every planter, whether he plants for the one, or the one thousand bales. And, having tested many experiments, I hope I may be of some service in drawing attention thereto.

The use of manures has been so fully set forth by our friend, Dr. Cloud, of Alabama, that it would seem a work of supererogation to allude thereto; yet, I may have some friends who would be guided by me, and as they might not be touched by the Doctor's able articles, I beg to say: As early as 1817, in Chester District, S. C., when boarding with my venerated friend, Mr. Harbinson, I saw the most marked difference as to yield of cotton, caused by manure, that I ever remember to have seen. In 1842 or '43, I tested stable manure and cotton seed as a manure; an unbelieving aged friend, as also quite a number of young planters, were called upon to express an opinion—it was, that a 500 lb. bale was the product; whereas, without manure, not one-half was growing, adjacent. I do not believe any manure can increase the yield of some of our western river bottoms, in that proportion; yet, upon thin land, I feel very certain that it can be done. I use cotton seed scattered in the drill, and then beds made thereupon, and always with a favorable result.

It is, I admit, a tedious process to haul out three or four hundred bushels of stable manure; but not less so is it to clear, fence, and break up new ground—nor more tedious than pulling up stakes, severing all the tender endearments of "mine own, my native land," to seek, at a heavy cost of time and money, a home in the western wilds, there to suffer from the combined attacks of mosquitoes and fever and ague! I have been on this spot for nineteen years—settled within twenty-five feet from where I now write, January, 1831—and have had some experience with bread from a steel mill, rolling logs, shaking with the ague, lost in a cane brake, and lying by the side of a log all night. I only say, to all dissatisfied with the old farmstead—go to work honestly, save all manures, plough deep, sow down peas, rest your land, and be a part of that land.

Another adjunct alluded to in the preceding paragraph—deep ploughing, and really breaking up the entire surface, not leaving any unbroken earth. Remember the mine that was left to a son, in his father's field! The youth thought that gold had been buried—he went to work with the spade, and dug up every inch, but no gold! He had to eat, and therefore planted—the product amazed him—he continued, and found the treasure—industry and good culture!

If the labor bestowed in California, with pickaxes, spades, &c., &c., was made available in the slaveholding States, I believe the mines of Golconda and the wealth of Crœsus, would fail to give an idea of the result.

And I fully believe that frequent, deep, and effectual ploughing, before planting, will do great good. One of the most successful planters in Hinds County, Miss., always bedded up and reversed his beds before planting.

I do not regard some of our large crop-masters as worthy of imitation—they make eight, ten, aye, twelve bales per hand, but it is by working negroes, and wasting land.

Alternation of crops has a powerful influence, and is of great benefit to the planter, if he will plant in four-fold rotation, cotton, corn, grain, and rest—absolute rest. I do not call it resting land to graze it. It would be as well to cut off the crop, and better, as the ground will not then be injured by trampling in wet weather.

Sowing peas (two to three pecks per acre) when corn is laid by, will give shade to the land, and a large amount of manure. Peas gather sustenance from the air as well as the land, and thus you return all to the land taken up by the pea, and more too. I am constrained to believe that a dense shade of pea vines will benefit land, even if every stem of the pea could be removed the first killing frost.

And last, though not least, I regard selecting of seed, duly curing before being bulked, as an important aid to the health and growth and productiveness of the crop. Why should not increase of vitality in cotton seed be beneficial, as well as sound and healthy parents to a sound issue? I admit I am interested in this being promulgated—but I hope not more than what all others should be. I am so well satisfied of the fact, that I have been purchasing seed for fifteen years or more. True, I have, in 1848-9, sold largely—but others are benefited as much as I have been. Yours, with respect,

M. W. PHILIPS.

SECTION V.—REPORT ON COTTON, TO THE NEWBURY AGRICULTURAL SOCIETY.

From the South Carolinian.

THE Committee on Cotton respectfully report: That the complete and thorough preparation of the soil is of the utmost importance to a successful cotton crop. As early, therefore, in the season, after the crop is gathered, as is practicable, the land should be broken up, deep and thoroughly. On clay soils it renders the ground loose and friable, and when stubble ground is planted, it is necessary to turn the sod early, in order to hasten the decomposition of the grasses and weeds. The beds should be made from three to five feet wide, according to the nature of the soil, and, if manuring is intended, they should be opened by a deep furrow, and the manure deposited and covered, by lapping on two good furrows. Unfermented compost manure, if applied early to clay soils, and buried with the plough, will be found most beneficial. It serves to keep the ground loose and friable, thereby giving access to heat, air and moisture, the three great agents of nutrition; and, in undergoing fermentation, it imparts warmth to the soil, which, in the early stage of the crop, is of the greatest importance to forward its growth. In making the compost heap, almost everything is available. Leaves and straw, which have been deposited in the stables and farm-yard, and ashes, should all be incorporated in the same heap. The compost recommended by the Committee on Manures at your last annual meeting, is an excellent preparation for this crop. By furnishing the hogs, when put up for fattening, with an abundant supply of litter, a large quantity of this excellent compost manure can easily be made.

When the same land is again planted in cotton, a deep and broad furrow should be opened with a large two-horse Eagle-plough, but if this plough is not available, then a substitute may be made by using a large twister, drawn by two horses, and passing up and down until the furrow is opened to a sufficient depth. In it should be deposited the cotton stalks, and a bed be made upon them by throwing up and completely covering them.

Early in March the beds should be prepared for planting, and to do this most effectually, a two-horse iron-toothed harrow, if passed over them, will reduce the land to a thorough state of pulverization. Follow this with a marker, making a small seed-furrow in the middle of the bed. The seed should be rolled, previous to planting, in a preparation of ashes, stable manure, and water, which is easily done, and embodies two distinct advantages. It enables them, when drilled, to assume a separate position, and also acts as a stimulant, which is very much required in the early stage of its growth. We feel confident, if this course of preparation were generally observed, we would hear less of bad stands, and the various complaints amongst our cotton planters every spring. The seed, when rolled, should be planted while moist, and immediately covered lightly. For this purpose we use a board, three feet long, slightly hollowed, which makes a clean sweep across the bed, leaving it in fine condition. A small harrow also answers a good purpose. As soon as a stand makes its appearance, the barring-plough should be run round, and followed immediately with the hoes, the crop cut down to the proper distance, say from fifteen to twenty inches, and the grass and weeds along the ridge destroyed. Then follow, with a suitable plough, at such a distance as will throw off the dirt from the cotton into the middle or water-furrow, thus covering up the grass and weeds. The hoes should immediately follow, leaving rather more than a full stand. The second and after ploughings should be more light, and, as the principal benefit which the crop derives is from keeping the crop loose, and free from grass and weeds; the expanding cultivator, in a season not too wet, will be found the best implement which can be used for this purpose. It will finish the row by running once through it, thus saving a good deal of labor in the crop, and the ground is left in a better condition than from the use of any other plough. Cotton is more benefited than any other crop by rapid working, and, in order to enable the planter to go over as often as his crop requires it, he should adopt the plan of only partially ploughing out the rows, and returning in a short time to finish the work. This, the use of the cultivator will enable him to do, on lands free from stumps and roots, but as it would be impracticable on soils where such obstructions existed, the use of broad shovel-ploughs and sweeps, applied in the manner recommended, would be of great gain to the crop. The cotton crop should be ploughed at least every ten days, and a furrow or two seems to keep this plant in as good growing condition, as if the entire row is regularly ploughed out. This season, a shovel-plough or a sweep would only have cultivated the grass, and those planters who used large turning-ploughs, and, by completely subverting the soil, smothered the grass, have succeeded in keeping their crops cleanest. In a dry year, the process of superficial ploughing, or scarification, by the use of the cultivator or sweeps, will answer an admirable purpose, and be the proper system of tillage, but in a wet season we have no fears in recommending the heavy turning-ploughs, for the economical cultivation of cotton. All this, however, is to be controlled by the season, and the preparation of the soil for the reception of the crop. Deep ploughing does no good in the middle of the rows, when the soil upon which the plants stand has not been broken up deep. Common sense points out to us that cotton, from its long, penetrating tap root, and the almost entire absence of lateral feeders, should be planted on soil of great depth of tilth, and, if this is furnished, we would open new feeding grounds to the plant, and thus, perhaps without the aid of manure on many soils, add to the amount of yield.

By the 1st of August all the cultivation of the crop should be finished, and from the 29th of July to the middle of August the cotton should be topped. This will cause the stalks to expand and perfect more bolls, and amply repay for the little labor which is consumed in effecting it. There are a great many varieties of cotton cultivated, and the Petit Gulf kinds are generally highly esteemed. Perhaps the McNutt is upon the whole to be preferred, as it bolls well, and is of early maturity; but it is generally known that all the varieties deteriorate after five or six years' culture, and it is therefore advisable to renew the seed, by introducing the most approved varieties from the Gulf region. The little expense which attends this amply repays the increased product of the crop. The Mastodon cotton was tried here last season under unfavorable circumstances. The piece of ground planted was wet, and as the seed was used sparingly, a bad stand was the result. The product, however, was found to be a remarkably good yield for the stand, and of early maturity, fully as early as the McNutt, which grew beside it. Should this variety, upon further trial this season, be found equally well adapted to our climate, as the other short staple cottons, it will add much to repay the planter for his labor.

The plan of cultivation adopted by Dr. Cloud* was tried here during the unfavorable season of 1845. The manure deposited was not, owing to the drought, available to the crop,

* See "South Carolinian" of Nov. 7, 1846.

but was of the utmost importance to the succeeding crop of corn. The experiment even then would have warranted its continuance, as by this system of manuring, the poor soils of our State would be annually improved, instead of being impoverished as they are, under the ruinous course generally pursued. Something should be done to restore the soil in some degree to its original fertility, and this can only be effected by a judicious and industrious course of manuring.

We know that there are many notions prevalent respecting a diminution of the cotton crop, and that the main objections to the cotton culture consist in the difficulty of continuing it extensively, and at the same time carrying out such a proper system of manuring and rotation, as will certainly and gradually improve our exhausted fields. This is an objection well founded, and if we were threatened with a dense population which would consume as much food as the land would produce, it would be high time to cast about for other systems of tillage. We are, however, not one of those who believe that the cotton culture is incompatible with the improvement of the soil, and instead of recommending our planters to decrease the number of their bales, we only go so far as to advise them to *produce a greater number upon fewer acres* than they now cultivate. When we reflect that the limits of our Society embrace the very best cotton region in South Carolina we are loth to see a culture decline which, in the days of past prosperity, has added so much wealth to our district. When cotton was first cultivated in this district, one bale to the acre was the average yield. The same amount can now by judicious manuring be produced on every acre of land which in its virgin fertility yielded that quantity. The labor of clearing an acre of forest, will always be sufficient to make an acre of worn-out land better than the new ground would be for the production of any crop, and an acre thus restored, would substantially add that amount of capital to the wealth of the country. The individual wealth of citizens has no beneficial tendency upon the people of the nation, the contrary being the general effect. To the general prosperity alone can we attribute national greatness.

When our small farmers become impressed with the necessity of cultivating their lands properly, and tilling them like gardens, they will soon render themselves independent of the fluctuations of the markets and the times, and to all such we say, *take your poorest acres, manure at least one, and, if you can, every one, plough deep*, and bestow as much labor on one-half the number of acres you now cultivate as you do upon all, and before a young man grows old, he will own a farm which will be a credit to his industry, and a rich legacy to his children.

All of which is respectfully submitted,

WM. SUMMER, *Chairman.*

Pomaria, S. C.

SECTION VI.—REPORT TO THE UNION (S. C.) AGRICULTURAL SOCIETY ON COTTON.

THE Committee on Cotton beg leave to make the following report: That the cultivation of cotton has so long engaged the attention of the country—aided by all the sagacity of interest—as one of the leading staples of the State, that it would be very difficult to advance anything new or instructive; but that the remarks which they propose to make, must necessarily be general and trite. They would remark in the first instance, on the importance of procuring the best seed. Experience has established the fact, that the quality of any article of produce may be improved by care in the selection of seed, in a climate congenial to its growth. It is a very general law of nature, that the offspring inherits the good or bad qualities of its parent; or, in other words, that like produces like. This is more particularly true, when applied to the vegetable kingdom. This section of country is not regarded as the best adapted to the growth of the cotton plant. The best efforts at selection from our own seed might not, therefore, be entirely successful, but still the Committee recommend it, as the best counter-agent against the degeneracy of climate—as the best means of preserving the seeds in their primitive purity. The Committee would therefore, also, earnestly recommend the frequent renewal of seed from the Petit Gulf countries. In the selection of seed from our cotton, it should be made from the second crop of, or middle bolls—the bolls on the lower limbs being generally small, and the seed more or less defective. The season for the growth of cotton is too short in this latitude to admit of its attaining to perfect maturity. Early planting is, therefore, very important. The planting should be so early, if practicable, as will just enable the young plant to escape the blight of frost. It is true, that sometimes late cotton, owing to the seasons, is most productive, but the chances are clearly against it.

The preparation of land for cotton, should claim the early attention of the farmer or planter, in several points of view. It increases the productiveness of the land, and consequently, the amount of the crop; it abridges the quantum of labor necessary to its proper cultivation; and it is requisite for the reception of the seed. The Committee deem it unnecessary to enter into the minutiæ of this subject, as both its importance and process are well understood, but will proceed to remark on the mode and manner of obtaining a good stand of cotton, without which it would be impossible to make a full crop. Besides the selection of seed, and preparation of land, before

3

adverted to, it is necessary that the seed be dropped regularly in the centre of the opening furrow, and with rather too profuse, than too sparing a hand; and then, covered carefully quite over, without displacement, with the soil, though not deep; using the harrow, or one-horse double bull-tongue plough for that purpose, after the Fairfield manner. If it is very dry, the latter is preferable, from the fact that it leaves a sharp ridge directly over the seed, thus preventing the ground from baking; and covering deeper, secures a sufficiency of moisture for germination, and, consequently, a good stand of cotton.

There is some diversity of opinion as to the distance that should be given to cotton. The greater the number of stalks that can stand on any given quantity of land, without interfering with each other, will make the maximum amount of cotton of which that land is capable. The Committee therefore suggest, that the rows of cotton be placed sufficiently close to quite cover the ground; or, in other words, that they be placed closer or wider, according to the strength of the ground. Thus, on poor land, the rows will increase so rapidly on any given numbers of acres, as to make them approach much nearer, in the amount of their production, to rich land, than is generally supposed.

The Committee will merely allude to the improvement of land generally, by manuring or otherwise, as being intimately connected with the growing of cotton—as that subject more properly comes within the province of another committee of the Society. They will only remark, that to grow cotton advantageously in this section of country, it must be done on fresh or manured lands. And thus, they have, in a very summary manner, disposed of the most important matters preliminary to the cultivation or the management of the cotton crop itself.

The Committee might, with propriety, rest their labors here, from the fact, that to say anything of the subsequent management of the crop is almost an act of supererogation. They, therefore, only propose to throw out a suggestion or so for consideration. A greater proportion of labor is bestowed on the cotton crop than any other, particularly on large plantations, where it is usual to plough four times, and close hoe at least three, and in some years oftener. It has occurred to the Committee that a good portion of this labor might be saved, by dispensing with close hoeing partially, or altogether, according to the field or year. The mode of proceeding is this: the hoe is made to follow the plough at the requisite distance, to ascertain what grass is killed by it—passing over all that portion of the row where the plough has done its work properly—and using the hoe only on the large bunches of grass, or such as may impede the growth of the cotton—thus relying on the plough to kill all the smaller grass. Some of the Committee have tried this plan of cultivation, and so far it has worked well.

It is a problem not properly solved, whether it is advisable to plough cotton approaching to maturation, or after it has attained to good size, with a full crop of *forms* on it. Your Committee think that to plough it, as in the case of corn after tasselling and silking, though it might not, with good seasons, and under favorable circumstances injure, but possibly improve it, yet it would be risking too much. On the whole, they are opposed to late ploughing under the scorching sun of August.

A variety of ploughs have been invented, with a view to their adaptation to the culture of cotton, but they have been pretty generally discarded. The scraper, or Eagle-plough, as it is sometimes called, is the best—answering some years and on some lands an excellent purpose—but its general utility is doubted. We should not, however, despond of others better adapted, when we consider the great inventive genius of the age. In conclusion, the Committee recommend a gradual reduction of the cotton crop, to such point as will admit of the improvement of lands, particularly—and generally, of improvement in such matters and things as appertain to the general economy of a farm. All of which is respectfully submitted.

JOHN GIST, *Chairman.*

CHAPTER II.

DR. N. B. CLOUD'S IMPROVED SYSTEM OF COTTON CULTURE.

SECTION I.—DR. CLOUD'S PLAN OF MANURING, PLANTING AND TENDING A COTTON CROP.

Extract from an article published in the Albany Cultivator.

THIS improvement when it shall have attained its highest state of perfection, contemplates the "*system of rotation*" in planting, under which the land designed for cotton lies the previous year in the state of fallow, which is found by experience most favorable to the growth and fruitfulness of the plant. I commence the preparatory operations for planting about the 1st of March, by spreading upon the land, *broadcast*, two to three hundred bushels of manure per acre—light stock yard and stable compost. I then run off the land in rows of three feet with a scooter-plough, opening a good furrow some three to four inches deep; this done, I take a large-sized shovel-plough and cross the scooter-furrows by rows, running at right angles of five feet wide. I am now prepared to commence manuring in the hill, having first ascertained that I have 2940 hills on each acre, which will require, by giving each hill a half gallon of manure—same kind of compost—184 bushels nearly, which I haul on the land in a cart, first graduated to a certain number of bushels, and with spades likewise prepared for the purpose, I deposit the requisite quantity of manure in each hill. By this means, which in practice will be found simple and expeditious enough, I give four to five hundred bushels of manure to each acre—*an infallible insurance* for 5000 lbs. of a superior staple per acre.

As the manure is placed in the hill by rows, the wide way, a short distance in advance, a *good plough-hand* follows with a turn-plough, which should run into the soil from *six to eight inches deep* at least, and turn well, with which four furrows are thrown together on each row; thus fixing the half gallon of manure in each hill, entirely within the *region of constant moisture.* This gives me a fine, large bed, and well broke, to lie until at or about the first of April, when the cotton seed should be planted. This is done by first opening the bed as shallow as possible, with some instrument such as that described by M. W. Philips, Esq., in the March number of the ninth volume of the *Cultivator.* This I prefer to any other instrument of the kind I have ever yet seen, since its depth of furrow may be graduated to a positive certainty so as to avoid disturbing the manure in the hill; it should not be opened out deeper than one inch. The bed thus opened, and the seed previously rolled in leeched ashes or sand, which answer very well, though I prefer a compound of two parts of ashes to one of common salt made moist with water; the seeds, well rolled in this, are carefully dropped over the manure. Eight or ten seeds in a place will answer to secure a stand. There will be no difficulty in dropping the seed over the manure in the hill, when it is recollected and observed that upon the unbroken space of some two feet between each row, the scooter-furrows will be found an unerring guide to the manure in the bed at distances of three feet. The seed thus dropped I prefer to have covered with a hoe, *lightly and carefully;* bearing in mind this golden truth, that "a crop well planted is half made."

Immediately after planting, the middles or unbroken balks should be ploughed out.

The crop of cotton thus planted, which should not exceed *three to four acres* to the hand, may be performed in good time and well done. In a few days, say nine to twelve, the cotton will be up, presenting a most healthy and thrifty appearance. The next operation to be performed, as early as possibly convenient, is to plough out the middles *well*, the wide way, with a good shovel-plough, having first run around the young plant with a scooter-plough. The hoe hands follow and thin the cotton down to two stalks, giving it a small quantity of soil. This operation *well done*, the plant is at once placed beyond all danger, since its tap root will now have taken such hold upon the manure below as to enable the plant to outstrip either grass or weeds, having yet to spring up.

Under this treatment, the *time-consuming* and worse than useless operations of *bar-shearing*, *scraping*, and *chopping out* are saved, as much to the benefit of the tender plant, as to the interest and economy of the planter, in despatching the *hurry and push* at this stage of the crop; and at the conclusion of this first working, I have my cotton growing off, and doing well. I have now no further use for a plough in its subsequent culture, but use the *sweep*—a kind of horse-hoe—I call it a sweep in the absence of a more appropriate name.

[Here follows a wood-cut representation of the sweep, a kind of plough used by some planters at the South. The one here recommended is made by welding two narrow wings over the point of a scooter, or bull-tongue, inclining backwards, with the ends of the wings two feet apart. It is so fixed upon the stock (that of a common shovel-plough) that it will not enter the ground deeper than one inch, if so deep.]

The great and singular advantages of the sweep over all instruments of the plough, harrow, or hoe kind that I have ever used, are these—that it will *kill* a greater quantity of grass and weeds in a given time, and *do less injury to the surface roots* of the plant, *so essential* to its *progressive* prosperity. The hoe-hands follow this instrument, thin the cotton to a stand, *one stalk* in a place, and draw up a small quantity of soil to the standing plant. The entire subsequent culture is performed with the sweep and hoe, which should simply scrape and pulverize the surface, so as to kill any grass and weeds that may appear, and allow a *free circulation* of atmospheric air to the fibrous roots of the fruiting stalk, requiring at this *critical period* all the aid and nourishment that culture, soil, and atmosphere can afford. By the 1st of July my cotton stands from five to nine feet high, and I have it topped by the 10th, at farthest, after which I run the sweep once through it, and the hoe, if necessary, to remove any grass that may have sprung up immediately above the stalk. After this, and by this time, frequently in places the cotton will be so much interlocked, and the ground so shaded, as to keep down all other vegetation ; yet it may be found necessary again to chop about in places with the hoe, when the cotton may not have locked so early. This should be invariably attended to. This brings us again to the season of harvesting the staple.

Let no planter prejudge and reject this system, upon the score of simplicity, supposing the process too simple to accomplish the object proposed; first, act wisely, make the experiment, and try it. Strictly follow this plain and simple process, and if the land does not reward your pains-taking, with *five or six-fold the quantity per acre*, of a *superior staple*, than has at any previous season been taken from it, in its natural state, I will present the experimenter with *one bushel* of my *improved seed*, with which to perfect the experiment. At another time I propose devoting a paragraph to the importance of *selecting and improving* cotton seed.

It will be observed that manuring constitutes a large item in this system of improvement, a source of revenue too much underrated by planters, and consequently too much neglected, because the subject requires a little extra attention—which attention is so essential to the prosperity and well-doing of a farm. Nor, gentlemen, have I seen anything better said, or more true, than I find in the sentiment, under the head of *a few queries*, in the last December *Cultivator*, where you remark to the planter and farmer, "In your manures is your *gold mine*, more valuable than any of the Carolina ones, and you should be anxious to increase them accordingly." But I hear some planters say, "It is impossible to produce so much manure;" this is, however, the result of inexperience, and the want of *determination*. I am entirely convinced, from my experience in making manure, that it is not only practicable, but a perfectly easy task to prepare, upon every plantation in the cotton region, great or small, 1500 *bushels* of an excellent article of compost, per annum, to the hand, at a cost of less than two cents per bushel, by the assistance of the stock of horses, cows and hogs, upon properly arranged lots. This is done by having the lots well littered, by throwing in pine straw, in large quantities and frequently, or oak leaves, where the pine straw is not to be had, with cotton and corn stalks, &c., and occasionally haul and scatter upon the litter a few loads of muck or marl, one or both of which may be found on or near every farm in the country; upon these lots, pen and feed your stock every night. The manure thus prepared, should be collected in pens or pits, three or four times during the year, after heavy falls of rain, and the lots replenished with pine straw, &c.; by this means a very large amount of manure is collected during the season, and that, too, at an inappreciable cost. Again, we have another difficulty. There are but few persons who believe that pine straw can be converted into manure; for the benefit and information of such, who may read this, permit me to quote a single sentence from Liebig's celebrated work upon Agricultural Chemistry: "The bark and foliage of oaks contain from six to nine per cent. of potash. The needles of firs and pines, *eight* per cent." But it is not on account of the potash exclusively, that I prefer pine straw, to all other vegetable matter, in the preparation of manure, since it possesses another invaluable quality, above all others, in absorbing the juices of the manure, which are thus saved from evaporation, and readily applied to the land. I doubt not but a single year's experience will convince every intelligent planter of the innumerable advantages of this improvement, and its perfect adaption to the culture of cotton and other crops.

I will now close this number by a very few remarks upon the character and quality of the soil upon which my experiments have been conducted. It is a high ridge-land, readily recognized, and its quality distinctly understood, in our southern country, under the name of *forked-teaf, black-jack, pine-barren,* a deep, porous, sandy, superstratum, lying under a tolerable good clay, at a distance of two to three feet below the surface. A true picture of nature, and naturally poor enough. This land, under the treatment above detailed, grew my cotton, from which I have gathered a greater nnmber of pounds per acre, (indeed, almost double,) that I hava ever seen recorded, is in its natural state, inferior to the average quality of cotton land, by at least one-half. I might refer you, if necessary, to more than one hundred gentlemen, planters from Georgia and Alabama, who have examined my experiments carefully, and several of them at various stages of its growth, and with one general consent, pronounced it a fair test, and a great improvement. I have, from several stalks that grew on the three acres, in the proper places, taken three and a-half to four pounds of cotton, *carefully weighed.* In the perfection of this improvement, yet in a state of great crudeness, when every stalk upon the acre (2940) *shall mature equally well,* what may I *reasonably* calculate to gather?

"——Nil desperandum,
Possunt quia, posse videntur."

N. B. CLOUD, M.D.

Planters' Retreat, Ala., Dec. 26, 1842.

SECTION II.—PRINCIPLES AND PHILOSOPHY OF THE IMPROVED CULTURE OF COTTON.

From the Albany Cultivator.

MESSRS. GAYLORD AND TUCKER:—Entertaining the profoundest respect and the kindest feelings towards the opinions and practices of those planters who are greatly my seniors in age and in agricultural experience, I propose, now, to engage in discussing "the principles and philosophy of this improvement in the culture of cotton." I will first remark directly, gentlemen, what I have intimated throughout this correspondence—that in conducting these experiments, and in advocating the claims of this improvement, (the leading and meritorious features of which belong to your invaluable *Cultivator,*) I have had no ambition to gratify which is not common to the lover of science and agricultural improvement; nor have I any interests to subserve thereby, which may not be the privilege of every planter in the country, however humble his pretensions or ability. Yet, admonished as I have been, by the precipitate and unmeasured tirade of vituperation and spleen which my first paper excited among the corps editorial of the country, and others of the "skinning gentry," I have not ventured upon this most delicate position, though long promised, without again revising carefully and practically, the principles and ability of this system.

In my first paper, it will be recollected, I stated the grand object which this system of improved culture proposes to accomplish for the planting interest of the country; and I also gave there a fair and impartial, and a most satisfactory expose of that system, (if such it may be styled,) by which the cotton plant is at present grown. In my second number, I gave in detail, in an equally plain and simple manner, the *modus operandi* by which my experiments were conducted. This was no second-hand report, or say-so of another person; but in part the work of my own hands, and the entire management under my immediate supervision. In this paper, I propose giving *the why and the wherefore* for all this—at least, in my humble opinion; and to point out the *inconsistent*, the *reckless* and *grassy* policy of the present practices of the country, as compared with the *systematic*, *economical* and *philosophic* policy of this improvement.

It is my purpose, gentlemen, first to give you a correct sketch of the botanic characteristics of the cotton plant, as we meet with it under the circumstance of its most favorable culture. I do not offer this in the spirit of ostentation, to appear learned from the technicals used, (the necessary peculiarity of every science.) My object is simply to call attention to a subject—too much neglected by planters—about which the books are carelessly in error; and a proper knowledge of which, in my opinion, tends greatly to indicate the best or right mode of culture.

The botanic name, then, of the plant we cultivate, is *Gossypium herbaceum;* we find it in the fifteenth class of the Linnean system, (*Monadelphia,*) and in the thirteenth order,

(*Polyandria.*) The first leaves that make their appearance after the cotyledons, three to six are *ovate*, and indicate certainly the sucker or branch limbs, that will first put out from the stalk. After these, we have the cotton leaf proper, *three to five lobed*, and mercronate, with one gland upon the *mid-rib* beneath; these leaves invariably indicate the arm limb, growing out horizontally, and articulated, forming at each joint one or more balls; coming out with the arm limb, we have almost invariably a branch limb, with several balls—or a stem, with one to three balls. The stalk is *ligno*-herbaceous and *pubescent*—and in our climate an annual, attaining the height of *six to ten feet.* The period of flowering is from 10th *June to frost;* the calix double, the outer one three-cleft; capsule three to five *celled,* with seven to nine *seeds* in a cell, involved in the staple. Early in the morning the *milk-white** bloom may be seen, in the form of a conic scroll, emerging from the fringe-work of its outer calix; and with the rising sun it unfolds the segments of its petal, and by one hour by sun we behold the full blown bell-formed flower. Thus, blooming white, it remains till twelve o'clock; when, within fifteen minutes thereafter, we may observe by the naked eye, a faint ray of pink skirting the thin margins of the segments, which pink color may be seen, by one to two o'clock, to have diffused itself throughout the bloom. It thus continues changing from white to red, till sun up the next morning, when it will be found a beautiful brilliant pink: now with the rising sun it gradually wilts, and by twelve o'clock it drops off, leaving a distinctly formed ball, securely sheltered by its calix.

This description, which is strictly correct, differs in several of its particulars from Eaton's, and from the miserably erroneous engraving and description of the same, to be found at page

* The Sea-Island Cotton bloom is yellow.

307 of "Sears' Wonders of the World; and yet, strange to say, this same engraving, with probably but a single correct feature, is copied into the *American Agriculturist,* in illustration of an article by Dr. Philips. I might point out a half dozen errors in that engraving; it will answer my present purpose, however, that I detain you with but two or three such notices. You will first observe the bloom, and the description given, and you will agree with me at once, that Mr. Sears has been *bugged* by an okra flower; the cotton bloom, in its healthy state, is never so much flared, nor has it any *red spots* in the bottom. Observe again, to the left, that young ball with its *drooping calix;* that is altogether unnatural, and is never seen except where the worm is or has been. You will observe the same error in the opening ball; every little boy, who has picked but fifty pounds of cotton, will tell you if that were the fact, there would be no trashy cotton. I am sure Dr. Philips has detected these blunders, with others equally evident. This, I suppose, will be considered a small matter, about which nobody is at fault; because even intelligent planters have never thought it worth while to give a correct description of the cotton-plant. This same carelessness is observed, when we cast our eyes upon the large map of Alabama; we there see a most imposing engraving of a large *fancy plant,* with its one hundred and one errors, if called a cotton plant. Observe again, the beautiful and chaste vignette of our own excellent and cherished *Southern Cultivator:* we see there an engraving designed for the cotton plant, yet I am sure if the *pendent* open balls were painted *red,* you would sooner take it for a pomegranate bush!

To the planter who is satisfied merely to plod along, the inanimate imitator of some skinning neighbor, this sketch will appear a tedious and uninteresting detail. I am convinced, however, of its importance—and there is a spirit of improvement abroad in the land, which requires just such detail of fact; because it is not possible, at least it is extremely improbable, that we succeed in improving and perfecting the culture of any article of vegetation, until we make ourselves *well* acquainted with its natural characteristics. Hence I remark, that when we look upon the stately pyramidal appearance of an improved cotton stalk, grown under favorable circumstances, we observe at once—indeed, we are forcibly struck with the distance proper in its arrangement upon the soil, which is so clearly indicated. Again, we observe an uncommonly large amount of foliage for an annual, besides some three to four pounds of seed-cotton upon the stalks—literally crowded from its base, upon an area of some fifteen to twenty square feet, to its apex, at the height of six feet. Now, in view of these clearly established facts, the invariable effect of certain well defined causes, I shall not suppose any planter so dull as not to know what course to pursue, if he find that it requires a given amount of grain to grow a pig to a given size in one year, that to produce another such pig the next season, the necessary amount of food or grain must be first supplied. Without the food, the pig will be found at the end of the year a *landpike;* and so the cotton, without the geine and manure, will be found, as is too common, the *little Frederick!* Were I to assume an affirmative position in this analogy, every planter would reply instanter, and most indignantly too—Sir, you are behind the times; our own sage Franklin, more than a half century ago, in his friendly advice to Poor Richard, has assured us, "that by constantly taking out of the meal tub and never putting in, we shall soon find the bottom." Philosophically true this—good homespun, and sound doctrine; yet plain and simple as be this doctrine, the cotton planter knows it only in song—his acquaintance with this golden truth is theoretic entirely. His exhausted fields and dwarfish, puny

cotton, tell tales more positively contradictory and gloomy, than I have room or inclination here to enumerate.

The governing principle, then, in this improvement, is to give constant and diligent attention to keep the *meal tub well supplied.* In the first place, produce and haul out upon your land a sufficiency of good manure, fully to supply the requirements of the plant all the season. In another place, I have shown that it is a perfectly easy matter to produce this manure, to which I will further add here, that the decaying materials abound spontaneously, scattered up and down, filling each nook and corner on every plantation, during all the season, a wasting nuisance that might be easily collected, and converted into a profitable revenue, if but one-third the time and attention, otherwise sedulously consumed in the butchery of the soil, in a petite war against grass and weeds, the inevitable produce of such latitudinarian systems of culture, was devoted to that most valuable employment. In this most important department of agriculture, science is actively engaged in rendering the planter the most essential service.

Having derived these important indications from the figure and natural characteristics of a perfectly matured cotton plant, the judgment of the planter is brought into active requisition in properly adjusting its relative position in width of row and its situation on the drill, in order that we secure the greatest possible advantage in its subsequent culture. My own experience inclines me to the opinion, that when land is improved only to the extent of one hundred and fifty to two hundred bushels of manure per acre, less than fifteen square superficial feet to each stalk will be too close. Nor will improvement carried to five times that extent, require greater distance than twenty square feet to each stalk. Since, then, it is found necessary that each stalk occupy this distance, it would appear that the simplest course would be to lay off the

rows equi-distant each way. The question is frequently asked, "Why not lay off the land four by four feet, or five by four feet?" There is a very serious objection to this simple plan, which every planter must perceive on a moment's reflection. In either case, the cotton will be found so entirely interlocked by the 20th of June to the 1st of July, as to forbid further work; yet we find, under the most favorable circumstances of seasons and culture, that it will take the stalk until the 10th of July to attain the height of six feet, short of which we should not top it, nor earlier in the season; and it is very desirable, and highly necessary even, that the cotton be swept once after topping it, which we find impracticable unless the rows be laid off wide one way, with a view to that desirable operation. Upon land, then, that is but moderately improved, I prefer the rows north and south five feet, by three feet east and west; and upon land in a higher state of improvement, six feet by forty inches will be found the best distance. Though we shall find the stalk a little crowded the narrow way by this course, yet we secure the more important advantage, in being able to scrape and pulverize the surface later in the season. I suppose there to be other advantages likewise, in this plan of laying the rows at right angles, north and south, and east and west, and bedding the land north and south; which, however, must form the subject of another article.

The next object to which I shall direct your attention, is the mode of culture which I conceive to be necessary in the after management of the cotton plant; the correctness, and even superiority of which, I hope to establish as clearly here, as in practice it has so triumphantly succeeded. The constant and invariable success which attends this improvement in my hands, is the result of a *strict and scrupulous adherence to system in its management.* Every science and every profession among men, which is either useful or valuable, acquires both respect and importance on account of system. System is essential to certain success in every undertaking; and especially is it necessary in this the first of all professions.

The principal object I have had in view, in all this manuring, thorough ploughing, laying off and bedding the land previous to planting the seed, has not been to plant alone; men plant abstractly, as handsomely and with the same facility, with less labor; it has been done to encourage and facilitate the early and extensive growth of the fibrous and soil roots, by which means the plant readily and equally early, augments the extent of surface (in number and length of its limbs,) for fruiting, and the consequent number of its organs of atmospheric nutrition. The immense advantages which the plant derives from an early accomplishment of an object so desirable, are at once obvious when we recollect "that the soil and atmosphere offer the same kind of nourishment to the roots and leaves of the plant." There can be no reasonable doubt, though I possess not the means of positive measurement, but that the plant multiplies its organs of atmospheric nutrition in precisely the same ratio that an improved and judicious system of culture facilitates the growth and prosperity of its roots. There is also another interesting consideration connected with this subject, which I esteem worthy of notice in this place; which is, that though the soil and atmosphere offer the same kind of nourishment to the roots and leaves of the plant, yet the character of its assimilation and consequent appropriation widely differ. My own opinion is, that the roots assimilate food for the production of the stems and leaves mainly, and that the leaves assimilate the same for the production and maturity of the blooms and fruit. I do not claim originality for this opinion; I think I have seen it hinted at in some work on vegetable physiology, though I cannot now say where.* I have been governed by this impression, at least in conducting my experiments, which have not as yet been of a sufficiently varied character to enable me to determine and assert the fact positively. My attention was first called to this interesting subject while investigating the cause and effect of rust upon the cotton plant, which every planter has seen, some of the features of which would seem to strengthen this position. How desirable is it, then, if all this be fact, that we adopt such system in our after management as will not only preserve this natural chain of action unimpaired, but encourage its progressive prosperity? It is not enough, however, that we thus dismiss this part of the subject; its importance requires of us a much more simple and extended view.

We will commence, then, at that age of the plant at which it is first worked, by examining the roots of two stalks; we pull up one in the ordinary way of thinning cotton, that is, we take hold of the stem and draw it up, and we have a single long root (in most instances), tapering to a point; we have simply the tap root. We will take up the other with a spade or hoe, the stalk standing in the centre of some six to eight inches square of soil, we then gently sift or shake the soil from the roots, and we have a fair specimen of the cotton root; we have what is properly meant by tap root, a plant with a main root long and tap-like, or tapering, dipping deep

* Since writing the above I see in a report of the sitting of the Academy of Sciences for August the 14th, a paper was received from M. Dutrochet, on the production and ripening of fruits. This gentleman states "that the removal of the leaves of fruit trees, in order to expose the fruit to the direct influence of the air and light, is exceedingly destructive." I suppose he means destructive to the fruit. If so, his experiments would seem to corroborate this opinion.

into the soil; besides this tap root, however, we find on almost innumerable quantity of fibrous or surface roots, diverging in every direction, as long, in many instances, as the tap root itself, and coming out, generally, from one-half to one inch below the surface. This is a fact worthy of notice, with which every planter may, if not already aware of it, acquaint himself early the next season. This may appear to some persons a very simple and a very trivial investigation, yet I find in it a most satisfactory solution of the immense injury which the cotton plant sustains, from the multifarious policy of the country. I remark, then, as the plant comes forward, so the tap root (where it exists, though an unnecessary appendage in our climate), sinks deep into the soil, while the fibrous or surface roots multiply and shoot in every direction; hence, I say, "as early as possibly convenient," after the plant is up, "plough out the middles *well*, the wide way, having first run around the plant with a scooter-plough." The main object in this operation, is once more, before the surface roots have come out so far as to sustain injury, to thoroughly loosen the soil, and again commingle it with the manure. The plant being now thinned down to two or three stalks in a place, and a small quantity of soil, molded about the hill, is left in this most favorable and growing situation. In the course of some fifteen to twenty days, when we return to work it again, it will be found to have come forward rapidly, standing from twelve to fifteen inches in height, and finely limbed. If we now take the trouble to examine a hill or stalk, we shall find an amount of earth included within the circuit of these fibrous and soil roots, as they penetrate all parts of the loamy mold, in pursuit of the luscious geine (like a flock of sheep fresh upon a rich pasture), that will weigh more than a hand can tote. With these facts before us, let us turn our attention for a moment to the practices of the country, at this stage of operations. One planter will now commence work, and the plant, standing from ten to twelve inches high, "with a bull-tongue or scooter-plough," and he will dagger into the soil, as close to the plant as he can possibly get, some three to four inches deep—he says, "to loosen up the earth, that the tap root may go down." Another planter will again, the second and third time, run the bar of a turn-plough to the cotton—he says, "to kill the grass;" thus it stands bedded in the middles, and "steaming" a few days, when these hot-beds are ploughed out; though I have even seen it barred the third time, before ploughing out the middles! All this may answer the *purpose fully*, and even look very well to the planter that operates to kill grass; but we have a latent cause operating destructively in this practice, and though the *certain effect* is not always willingly recognized, in the turning yellow and falling leaves of the plant, it is not, however, the less obvious. The planters operating thus will tell you, in the first instance, "this cotton has received a fine working; there's not a sprig of grass or weeds to be seen; but it does not grow off as it should; this *little dry spell* has checked its growth." But partial showers may have fallen upon the other man's cotton; he says, "See my cotton; how clean and nice it is worked, though it is too wet, and does not grow; *rainy weather* does not suit cotton." This is the logic (I will not say universal) of the devotees of this *grass-killing* policy, in accounting for its disastrous consequences, and will, I am sure, be very readily recognized as such by every impartial man. Now, the truth is, I will illustrate the whole difficulty here, by a very simple, though rather uncouth simile; it is, however, not the less pertinent to my present purpose, because men are not to be benefited, nor will they improve in the practice of any science or profession, unless they exercise the faculties of thinking and reasoning, though such exercise be bought at the expense of decent ridicule. We will suppose the planter operation in this was, having received a pair of fine Berkshire pigs, says to his trusty man, Sambo, "take this bushel of corn to the barn-yard, and feed those pigs well; I want them to grow *large* and *fat.*" Well, Sambo, always anxious to carry out the views of his master, and having carefully watched his operations in the treatment of his cotton, to make it *grow large*, takes up the basket, and then, providing himself with a hammer or hatchet, he proceeds to the yard; he first takes hold upon the gentle, unsuspecting grunters, one by one, and, with his instrument, knocks or breaks out their teeth, and then, throwing down the corn, he returns to the house with spirits buoyant, in the consciousness of having so consistently discharged his duty! "Well, Sambo, you have given those pigs plenty of corn, ha?" "Yes, sir, they are well fed." In a few days he takes a friend to look at the fine Berkshires. Yes, Sambo has given them corn plentifully; there it lies by them! But this is too plain a case; the pigs have the *teeth-ache, and they are broken off, too!* neither the wet nor the dry weather has caused the mischief here! And yet the pigs, like the cotton, have only *their teeth broken off!!* Poor Sambo! we leave him to explain to the world the rationale of this *root and teeth-breaking policy.*

N. B. CLOUD, M.D.

Macon Co., Ala., Nov. 1, 1843.

SECTION III.—EXPERIMENTS IN MANURING COTTON.

GOV. BROOME:—Immediately on ascertaining the result of my first extensive experiment in manuring and spacing cotton, I communicated the facts to the *Albany Cultivator*, an agricultural paper that had, at that time, quite a large circulation in the cotton-growing States. My object, as expressed at the time, was to have these experiments tested in various sections of the cotton-growing region. I gave the details carefully and minutely. I saw all the difficulties, and feared the *result* in the hands of gentlemen less interested than myself. The great principle of the improvement was a fixed fact. The extraordinary yield of cotton, the small area of land, naturally very poor, occupied in its production, and the *home means* employed, were facts too striking, and of too much importance, to be overlooked or slightly regarded by me.

As I have stated previously, it mattered not in a "first, crude experiment," what amount of personal trouble it gave me to so adjust and arrange these *home means* to produce such extraordinary results. The greatest difficulty connected with this experiment, was the trouble in getting a stand,—next to an impossibility. I had never seen manure applied to crops in any other way than in the hill, which succeeds finely with corn, but with cotton it is entirely different. Where the manure applied in the hills for cotton is worth the labor of application, and enough is used to produce a decided benefit, one half the hills at least, will either fail to come up or die immediately after coming up. This is an inherent difficulty in the plant itself, from its mode of germination, which I ascertained during the three succeeding years that I devoted to the subject for the express purpose of overcoming this main difficulty. The cotton seed, in the process of germination, attracts from the surrounding soil, and from the atmosphere, an unusual amount of water, as compared with other seed undergoing this process. Any artificial condition of the soil, which concentrates immediately about the cotton seed at this time an undue quantity of alkaline, gaseous matter, causes this fluid, contained in the tender, reticulated, or mesh-like incipient vegetable fibre, to undergo a species of fermentation, which of course destroys the vitality of the young plant. The plant is subject to this influence where a remunerative quantity of good manure, either compost, guano, or chemicals of any kind, has been used in the hill, even after having put out the third and fourth leaves.

Whether philosophically explained or not, the discovery of the fact cost me three years of the closest investigation. The tap root of the cotton plant does not make its way into the soil a perfectly organized root; the sprout which is the root, leaving the seed at the small end, dips directly downward, where it commences pouring out a semi-fluid substance, which is attracted downward partly by gravitation, and partly, perhaps, by electricity. This substance, like a small streak of smoke, is remarkably fragile, constantly and rapidly descending. It is the mould in which the tap root is formed. Any person who will take the trouble, can ascertain this fact for himself. Thus it is easy to understand how it is, that an unnatural, alkalization of the soil in the immediate vicinity of such condition of vegetable existence, should affect its vitality.

At the end of the third year, I determined upon a new mode of application entirely, which consisted in spreading all the manure used broadcast. This was done by hauling the manure out on the land and depositing in heap rows, say thirty feet apart, and the heaps thirty feet apart in the rows, with ten bushels of manure in each heap. The cotton-rows being first laid, the manure was spread broadcast, and the land bedded out. On or about the 10th of April, the cotton seeds were planted after a spacer, by which the hills are regulated precisely as desired. The result was a perfect stand, with the cotton healthy, and all of the same age. There is no difficulty in understanding the difference here in favor of broadcasting the manure, and in bedding out the rows. It is not deposited, a half-gallon in a place, but is incorporated evenly and uniformly throughout all the soil. The consequence is, that however rich the manure may be in alkaline matter, its thorough incorporation with the soil, so quickly and effectively dilutes it, as to render it entirely inoxuous to young cotton. There was no part of the experiment that gave me so much satisfaction as this. Every planter knows the value of a first, uniform and perfect stand. I use the term perfect, because by the use of the spacer, I approximate nearer a perfect stand than it is possible to accomplish by any other process.

From the interest and close attention bestowed upon this subject in all its various relations, the season had not expired before I clearly saw—as I then thought, and as subsequent experiment has and still is demonstrating,—a grand system of plantation economy, destined to revolutionize entirely the petty land-wasting customs of the country. You nor I, my very dear sir, may never live to see the day when that very *last man* shall cease to lay his cotton-rows up one hill and down the other, thus draining off the vitality of his land every three to three and a-half feet, to the depth of his puny plough, or to waste the sure means of keeping up the fertility of his fields, by feeding his stock in the public roads, or on the branch-side; but, with the lights of science and experience before us, wisdom clearly points out the course which it is our duty to pursue while we do live, from a three-fold binding consideration: first, our individual interest; then the true interest of our country; and, lastly, the obligation we are under to the true interests of our children, to use diligently every means in our power, to inform ourselves and the public mind as to the most economical modes of plantation economy. I have no patience with the inactive, inoperative friendship for agricultural improvement, of those clever gentlemen who tell me continually: Sir, your systems are beautiful, your exertions are praiseworthy, all your manure and manure-making, with your grade-ditching and horizontaling, and your rotations, &c., &c., are conditions actually essential to the improvement of our agriculture; but,—say they,—like every other country, this beautiful forest must be felled by the ruthless hand of Mr. Carenot, all this maiden and fertile soil must first be exhausted and washed into the branches, gurgling in pure and limpid water from the hand of nature, and the fields defaced by gullies and poverty-grass, and not till then can we give in to a complete and perfect system of improvement.

I beg to be distinctly understood here, as alluding to the great principles of improvement, and not to any individual practice under it. In my own practice and system of rotation, which I have had in successful operation here at La Place since 1846, I am not immovably confident that I have hit upon that arrangement under the principle that is to accomplish the best results. So sanguine am I, however, that it is worthy of general adoption in its main features, that I feel no hesitation in commending it to the consideration of those planters who have determined to begin the good work of improvement. As a matter of course, the circumstances of locality will, to some more or less extent, modify the practice; but the principle remains the same.

Having thus disposed of the Experiments, I shall, in several subsequent numbers, treat the subject as a matured system of plantation economy; showing, as I think, and, as my practice clearly proves, the eminent advantage of a proper rotation, even in cotton planting. In doing this, I shall respond to your various inquiries of stock, stock-feeding and manure-making, &c., as they come in place.

DR. CLOUD.

ECTION IV.—SYSTEM AND ROTATION IN COTTON CULTURE.

Gov. Broome:—I propose, in this article, to detail that system of rotation and shift of crops which I have in successful operation here at La Place, and which has thus far given entire satisfaction. In adjusting and adopting this arrangement, I have not been governed so much by the largest amount of cotton that might be grown on the plantation, as by the amount of independence in plantation economy, which the capacity of the farm, under proper management, is competent to secure to the labor and pains-taking of the proprietor. In other words, after innumerable experiments and tests, this system has been adopted as the one best and surest, calculated to feed and clothe the operatives of the plantation, supply all the stock necessary to its various uses, improve annually and protect the fertility of the land, and leave, at the end of each year, the proceeds of a fair cotton crop as the clear profits of the plantation with all its outfit. I shall not presume to say that there have not been favored localities in the older planting States east of this, whereon three of the above-stated important conditions of plantation independence were for a time possessed; nor do I say that there are not such favored localities in the new or western States; but this I will say, that the total absence and disregard of the fourth and all-important condition, the improvement and protection of the fertility of the soil, together with the increasing population of the country, having shorn such favored localities in the old States of these advantages, will deprive them in the new States, wherever the great principle of improvement is disregarded, in the absence of some system of plantation economy that might otherwise sustain them. It is this great error, this fatal error in the plantation economy of the cotton-growing States, I have diligently, for fifteen years, sought a remedy for. I have at no time been interested to teach planters how to make large crops of cotton and corn on rich land. I do not know an industrious man in Macon county, who cannot grow a large crop of cotton and corn if he has rich land to cultivate. Sambo, with no other instruction but the observation gathered from the hurried directions of his overseer, can, and frequently has, on rich land, made a big crop of cotton. And it is in this phase of the question that this fatal error is seen in its strongest light. Look back, if you please, toward the rising sun, and see the scant pittance with which land, once rich in its maiden fertility, now rewards the industrious labor of the *merely plougher and hoer.*

My chief object has been, in patiently prosecuting these experiments, and in watching and investigating their results, to devise a system of plantation economy which, while it will, in the aggregate, bountifully remunerate the industrious labor and pains-taking of the planter, will, at the same time, make poor land rich, and rich land better. The allurements of an honorable and lucrative profession, and the jibes with the pointing finger of ridicule from kind friends, have proved equally unavailing in diverting my attention for a moment from the one great object; and I may now exclaim, and do, triumphantly, *Eureka!*—I have found it! And if there be a single feature about this system that affords me more pleasure than another, it is, that the perfection of the system, with all its advantages, are as accessible to the planter of humble means, as to the planter of more extended means. There is nothing foreign, intricate, or costly about it. It is the production of the country, the soil, and the climate where we live.

It is immaterial what number of hands may work on the place, we allot to each twenty acres, and upon this condition proceed to divide the land into four equal parts, adopting the system of four years' shift as best suited to our plantation economy. The first object which I direct attention to, is to grade —ditch the land where necessary (which it is generally), and horizontal the rows perfectly level—this is proper and superior to all other plans on sandy land. In the next place, I fix the rotation, and shift thus: five acres to each hand in cotton, ten acres for grain, and five to lie in fallow. Our system of shifting crops proceeds in this way. I plant cotton on the same land once in four years, and the cotton is always planted on fallow land, with a dressing of 500 bushels of compost or stock yard manure per acre, which is spread on the land broad-cast, and incorporated with the soil uniformly in the process of bedding out the rows. This will be more minutely explained under the head of "Application of Manures." Let it be borne in mind now, that this land is perfectly level, and that all rain water sinks into the soil where it falls, and the residue of the cotton stalks, leaves, burs, blooms and limbs, with the seed, except for planting, are all returned back to the same land where they grew. Upon this land the next year we plant corn, manuring it with cotton seed. But to our corn crop, which I regard as the most important crop on the plantation, we add two acres of the land which was in corn last year, thus giving us seven acres in corn to each hand. On the other three acres of that portion that was in corn last year, we sow small grain, which upon land thus treated, will furnish a sufficiency of oats, rye, and wheat, for the wants of the plantation, when you have such a crop of corn as we provide for. Then we have lying in fallow, for the next year's cotton crop, the three acres that were in small grain last year. Every one will see at once the simplicity of this system of rotation and shift of crops.

I will now endeavor, as briefly as possible, to give the reasons why I believe this to be the best system of rotation and shift of crops that can be adopted in a cotton-growing country. In the first place, it embraces all the conditions necessary to sustain the cotton-planting interest within itself, independent of external or foreign aid. To this feature, I think, there cannot be too much importance attached. Again, the several crops succeed each other to better advantage, both as to their culture and healthy growth, than in any other way that we have seen or attempted. It may not be generally understood by planters from practice, because it is not a common practice, indeed it is of the rarest occurrence, how well cotton grows after one year's rest or fallow. I conceive it to be, in its healthy, vigorous growth, and exemption from insects, more like growing cotton on fresh land. Nor will this be difficult for any planter to comprehend, when he recollects that on the fallow I spread 500 bushels per acre of good stock-yard compost, or its equivalent.

I am sure I shall have no difficulty in persuading any planter that corn grows better, bears better, and is less trouble to cultivate after cotton, than after any other crop. So well, indeed, does it do, after a crop of cotton that has received a dressing of 500 bushels per acre of manure, that it is yet a matter of uncertainty with me, after twelve years' experience, whether or not a good corn crop is not more certain without than with the seed ; and if we have drought, it is certainly best not to use the seed on corn thus treated. Then we have the seed to add to our compost heap for our cotton. Then, again, the effect of the corn and small grain crops on the land being about the same, I prefer placing the small grain after the corn, as it does better after corn than corn does after it. After the small grain, the land lies one year in fallow. I have a theory about this four years' shift and one year in fallow, in regard to its curative influence upon the diseases of the cotton plant. Of course I cannot go into its explanation here, but I give it as my opinion, that if the same land throughout the country was planted in cotton but once in four years, it would prevent the insect of rust—I am sure it would of lice, and I think it altogether probable it would do much toward relieving it from the injury of the bole worm.

Under this treatment the plantation is every year improving. From the extent of pasturage which it affords, and the large amount of corn raised on the plantation, an average of 250 bushels per hand, there will be no manner of difficulty in raising all the stock, hogs, mules and cattle, that are needed on the plantation. It has been objected to this system, that in the extent of pasturage afforded, prairie and clay land would become too much trod by the stock, causing such land to run together and break up clody. I am confident the objection is unfounded, as the great object of the system is to accumulate on the land the largest possible amount of vegetable matter, which, while it keeps the land loose and friable, contributes so largely to the luxuriant and healthy growth of cotton. These objections, that fail to stand the theory and science of agriculture, fall to the ground as impotent and futile, when we examine the same system (in principle) in successful practice in the States of Tennessee, Kentucky, Ohio, &c., on calcareous clay lands, raising by pasturage, &c., not only mules, horses, hogs, and cattle for home consumption, but *for all our cotton planters*. There is an incompatability here certainly. The only precaution necessary, is to prevent stock running on the land while wet with rain water standing on it.

There is nothing more easy than to account for this false alarm among cotton planters. See the sedulous care, if you please, with which they have drained the vegetable strata of their fields, for the last forty years; each row is a perfect drain, not of water alone, but of vegetable mold, the life's blood of the land; the cotton and corn stalks generally burned; thus denuded and leached, it is not surprising that the hoof of a hungry cow should poison it!

It is further objected, by those otherwise approving the system, that it will not make cotton enough; that it does not lot sufficient land to secure every year a full crop of cotton. To this objection we simply oppose at first this fact. No man in this country, on the same quality of land, has realized from 1844 to 1853 inclusive, to the same proportion of hands, what I have, notwithstanding I have been experimenting all the time. If I have not made as many bales of cotton, which is improbable, I have raised that which cotton had to supply necessarily. This is obvious in the substantial improvements on the land, and its increased value, at least five hundred per cent.; not that I could simply sell it for that much over and above its cost twelve years ago, but it is its absolute annual production. Nor does it possess any artificial advantages of railroad or city value, as land in sight of it of the same quality, and just as valuable in 1843, under the "kill and cripple policy" of the country, sold last year at less than $6.25 per acre.

DR. CLOUD.

SECTION V.—SYSTEM AND ROTATION IN COTTON CULTURE—CONTINUED.

Gov. Broome:—In the Nov. number of this Journal, *(Am. Cotton Planter,)* I did not conclude all that I had to say under this head. I detailed there the "System of Rotation and Shift of Crops" that I pursue here, and in which I have the fullest confidence. The object of this article is to show that some such article as this, producing the same results is essential to the renovation of our already exhausted fields—to retain and improve the productive quality of our new lands, and to secure at the same time the raising at home of sufficient provisions with plantation teams, enabling us entirely within ourselves to carry forward the prosperous production of our cotton. In every other section of this country, north, east, and west, the proceeds of the productive industry of the people in the grand aggregate, are retained at home, while we, the planters of the south, producing annually, from a single one of our crops, $150,000,000 ! pay out the grand aggregate to others for bread, bacon, and mules, all of which we may, under a proper system of plantation economy grow at home, and thus we may retain at home also this large sum of gold, the substance of our fields, to be expended in home improvements.

It is an entirely fallacious political economy that supposes for a moment, that we are to make so much cotton annually, at the sacrifice of our personal and national interests; and it is as equally fallacious to argue, as many do, that it is our true policy to buy bread, bacon, and mules of others—though we may be able to raise them—that they may be induced to buy our cotton. There are other arguments for this ruinous policy too frivolous with which to detain you.

Now I insist upon it boldly, that this whole barter policy is totally at fault. It is one of dependence and slavishness. With a climate and soil peculiarly adapted to the production of cotton, our country is also equally favorable to the production of all the necessary cereals, and as remarkably favorable to the perfect development of the animal economy, in fine horses, fine active mules, good milch cattle, superior sheep, and fat hogs, and for fruit of every variety (not tropical) it is eminently superior. If this condition of things be fact, and I assert it to be such, why is it that we find so many *wealthy*

4*

cotton planters, whose riches consist entirely of their slaves and worn out plantations? I desire to show, and I shall prove it in practice, that a judiciously arranged system of plantation economy will secure upon the plantation sufficient grain, bacon, and mules to supply its wants; and a cotton crop, unincumbered by these absolute necessaries, that realize a handsome dividend upon the capital and labor of the planter. In this cycle of rotation and shift of crops that I practice, there is afforded, in the first place, every necessary means of improving the fertility of the land. Another striking feature about it, and not the least recommendatory of it, is the amount of rich pasturage that it affords for stock. I regard this as among its highest recommendations. Stock cannot be raised successfully or advantageously without pasturage, in addition to well-filled cribs of grain. The quantity of land appropriated under this arrangement to corn, secures a sufficiency of that grain for all needful purposes. This crop should always be laid by early, and peas, the common cow-pea, or some of its varieties, sowed broad-cast over the land and ploughed or harrowed in, which adds very materially to the value of the pasturage, as well as improves the condition of the land. It is argued by planters generally that grazing land injures it more than the stock are benefited by the pasturage. The argument is too often illegitimate! The land is first ruined by the one-crop practice of cotton, &c., till the vegetable mold and inorganic salts of the surface and ploughed soil are exhausted, it is then turned out to pasture. It soon runs together, of course, produces little grass, and sustains poor stock. The difficulty is not so much in the injury, which the hungry stock did in grazing the pasture, as the ruinous system of culture that prevented any pasture at all. Land under an improving system of culture is not thus affected. Rich land upon which water is not permitted to run, whether naturally rich or made so by art, furnishes a wilderness of grazing, when turned to pasturage, which not only greatly improves the condition of the stock, but retains a sufficiency of refuse vegetable matter, which, after the plough, keeps up the loose and friable condition of the land. It is in this view of the subject, that we see this self-sustaining system of plantation economy. Under this system, or any one like it, furnishing the amount and value of pasturage that it does, the raising and keeping of stock, mules, hogs, and cattle, necessary to supply the wants of the plantation, become a source of absolute profit—the land is made rich, and continues improving in the production of the elements of fertility—the compost manure is made valuable, because it is trod up and mixed with the excrements of stock kept fat on rich pasturage. This rich compost manure, applied to the land once every four years, in quantities sufficient to make a bale of cotton per acre, continues to improve the land and thus increase annually the grain crops and pasturage. All this is simple, plain, and practical.

It is objected to this country by planters and others taking their cue from them on account of its "short bite" and sterile pasturage, as they are pleased to call it. Nor has there been a designed misrepresentation in this: it is the result of observation derived from the working of this universally draining system of growing cotton. Now the facts which my practice and observation under this system have demonstrated, are these: that no country is equal to this for good and "*long-nip*" pasturage! Our climate is remarkably favorable to rich and luxuriant pasturage. The red man of the forest and the pioneer white man, that came here in advance of our "scratching ploughs," tell us they found the wild oat and native grasses waving thick, as high as a man's head, and so entwined with the wild pea vine, as to make it difficult to ride among it, all over this country. Every cotton planter has heard of these fine primitive pasture ranges, and many have seen them. If the country or the climate has been cursed in our appearance as planters here, it has been in the wasting system that we introduced and continue to practice. There is no grass, for hay or pasturage, superior to our crab grass, a native to the "manor born." Up by the 1st of April and continues green and growing (when properly managed) throughout the summer and fall till frost. The land once set with it never requires seeding again. Our crow foot is also a most invaluable late summer and fall grass. The short and extreme mildness of our winters, with the various evergreen or winter grasses, in connection with red clover, rye and barley for winter and early spring grazing, enable us to keep stock through the winter cheaper than farmers can in higher latitudes.

Under a system affording such facilities for grain in abundance, rich and extensive pasturage with fat home-raised stock of every variety and land improving annually in fertility, the culture of cotton becomes a process of gardening, productive and remunerating. The land may always be wrought to the best advantage, without injury at any time to either crop or soil.

Again: cotton thus treated matures earlier, feeds and fruits more rapidly, being strong and healthy, and less affected by insects, lice, rust, or the worm. Of course, then, it opens earlier and may be gathered to better advantage and in better order. It also affords a greater degree of certainty for a fair crop, both to the land and hand. This is the result of causes, both legitimate and philosophical; first, the land is provided with the food in proper form and quantity, which the cotton plant requires to bring it early to maturity; again, there is time and opportunity afforded to prepare the land for the reception of the seed, and the mode of seeding also secures a stand, perfect, regular, and uniform throughout; by perfect I mean the mathematical arrangement by which the hills or stalks of cotton are so placed on the land as to feed equally, grow uniformly, and at maturity, fill up the land completely.

In the January number of this Journal, we shall treat fully of the preparation and application of plantation compost manure, with some remarks perhaps on the application of guano and its value as a fertilizer in southern agriculture, the result of some twelve years' experience.

DR. CLOUD.

SECTION VI.—COMPOST MANURES; STOCK-YARDS, ETC.

Gov. BROOME:—The preparation of stock-yard compost manure, and its proper application to the soil, as a fertilizer, in the production of our important crops, cotton and grain—with some remarks on the value of guano to the Southern planter, will claim our attention at this time. This species of fertilizer, the most common, and cheapest to the planter, is valuable in proportion to the care and attention exercised by the proprietor in its preparation. This fact I have clearly shown in a previous article. I have given this subject much careful attention, and I am thoroughly convinced that too much importance cannot be attached to it, as an integral item in our plantation economy. Compost manuring, in connection with stock raising and pasturage, is the true renovator of all agricultural exhaustion. Stock are the inseparable companions of agriculture. All the team service of the plantation they perform. They also furnish quite a considerable proportion of the food consumed by the family and operatives of the plantation. In the performance of all this important service, they must consume, on their part, a very considerable proportion of the produce of the plantation. In this consumption, however, of hay, fodder and grain, under proper management there is nothing really destroyed or lost to the plantation. It is at this point the great difficulty is encountered by planters, in the preparation of compost manures. When the range is relied on for stock raising and feeding, as is almost universally the case in the planting States, the penning and shelter of stock every night is attended with a great deal of trouble, and the food consumed —after the first month or so in the early spring—is of such a character, and procured at such toil on the part of the stock, as merely to sustain animal life, and their excrements, of course, almost valueless as a fertilizer—at least comparatively so. This fact, connected with the rude and careless means usually adopted on plantations for composting and saving manure, furnishes the criteria upon which the opinion of the planting public is based, as to the value of compost manures, and the importance of its preparation in the plantation economy of the country.

In an article published in the November number of this journal, extracted from a premium Essay prepared for the "Maryland Agricultural Society," the position is taken that compost manures are not worth the hauling. This is the result of experience in Virginia. This opinion is very common all over the country, and it is the effect of that state of things which we have detailed above. My experience for the last twelve years, has led me to a very different conclusion. An alysis shows that the dung of animals—the horse, cow and hog—well kept, abounds in the very same fertilizing elements that make guano so valuable. If, then, the proper treatment of stock on the plantation fit them for the greatest value as teamsters, milkers and porkers, and in that condition their excrements produce the most valuable fertilizer, how important is it, in an agricultural point of view, that the fact be distinctly understood and acted on by the planters of the country. My experience fully sustains this position. In a previous article I have shown that this system of rotation and shift of crops furnish the necessary means, in rich pasturage and abundance of grain, to keep the stock of the plantation in proper condition. In this condition of the stock of the plantation, I may answer another one of your inquiries, as to the number of stock that may be thus kept to the hand. This answer is properly in place here, previous to entering upon the details of preparing compost manure. Twenty head of cattle to five hands, will answer all the wants of the plantation. The number of hogs is to be measured by the bacon necessary to do the place. Plough teams, one for every two hands, and sheep enough to clothe the negroes. Of course, on large plantations the exact number cannot perhaps be preserved, but about this proportion will be found to answer every needful purpose. Now then, on a plantation thus arranged and stocked, as mine is, I shall proceed to give in detail the plan of operations which I pursue, by which I am enabled to make 2,500 bushels of good rich compost manure per hand every year, and the only proper mode of applying it to the land.

In the first place, the farmer's golden rule is emphatically applicable here, and, I may add, entirely essential to success—"a place for every thing, and every thing in its place." Each kind of stock must be provided with lots and shelter, and they must be induced or driven into their quarters every night during the entire year. These lots, stables and shelters, are to be constantly and regularly kept well littered with vegetable matter, which being broken and tread up by the stock walking and trampling over it, forms a most valuable absorbent for preserving the fluid portions of the excrements. For gathering pine straw, oak leaves, and other decaying vegetable matter from the forest, I have seen various plans recommended, such as detailing such hand or hands and cart, for every five or ten hands on the place, &c. But I have found no plan to answer so well in practice as this: I have prepared for each hand a good, substantial and handy iron-toothed rake; during wet, rainy weather, all hands with these rakes gather rapidly large quantities of vegetable matter, which is as readily hauled into the lots on large frames made for the purpose. This is a general rule, and rigidly persevered in during all the year, except in winter after the crop is gathered, when I have it hauled into the lots as it may be needed, as we are not then so particularly engaged in the plantation. In the spring and summer, after every fall of rain, all hands are engaged in raking up and hauling litter into the stock lots. Under this arrangement, a day after the fall of a wetting rain can be more valuably employed by the hands of the plantation in collecting materials for preparing manure, than by ploughing or hoeing the wet soil. Every planter knows well the injury done to the land by working on it while wet. The crop is not benefited by work done at such time, nor is the grass or weeds so likely to be subdued. But the time may be most valuably employed in preparing the materials for composition manure, and when the land is in proper condition for work, the cultivation of the crop is resumed under the most favorable circumstances. The great point gained is this: the large amount of rich, productive manure, which being applied to the land, under judicious culture secures the production of the desired crops on one-third the surface required on the same land to grow it without the manure. After the preparation and planting, manured land being just as easy to cultivate as that unmanured, the time for preparing manure while the land is wet after a recent fall of rain, is most profitably employed. All decaying vegetable matter about the plantation, such as weeds, grass, &c., that grow and collect in the fence jams, in low wet places, in the ditches, &c., should be carefully raked up, and at a convenient time hauled into the stock lots. Muck, also, where it may exist in ponds and branches within or contiguous to the plantation, should be hauled up in the summer while dry and light, as nothing contributes more valuably to the compost heap, nor is any absorbent perhaps more retentive of the valuable fluid portions of fat animal excrements. This is the process by which I am enabled to prepare the large quantities of rich, valuable compost manure per hand, which I apply to my land annually.

There is another important item in the preparation of manure, which should be mentioned here. It is the construction of the stock lots. This should be done in such manner as to prevent any water from running into them, that does not fall immediately on them, nor should any water be allowed to escape from them. Moisture is a component part of compost manure. Too much water, however, adds more to the expense of carriage than to value in fertility. This teaches the economy of housing and sheltering the compost heap, that we may be spared the expense of hauling to the field so much water, quite as heavy as the manure itself, and of no value. Of course, every planter engaging in the preparation and saving of compost manure, will consult the conveniences of locality, &c., of his plantation, in the construction of his stock houses and lots, and other arrangements for the business.

I shall now give you my mode of applying the manure to the land. Of course I esteem it the proper mode. As I have stated elsewhere, my land, though but little undulating, is all laid off in rows, as nearly level as instrumental operations can accomplish. The manure is hauled out on the land in carts, in tumbling bodies, graduated to hold an exact number of bushels. In the commencement, a row is selected, fifteen feet from the fence or beginning. This is the heap row. Fifteen feet from the end of this row, the first heap, or half the load is deposited; it is raked out by removing the hind gate of the body. Thirty feet from this, on the same row, the second heap is made by tumbling the body, when all the manure slips out, and the further trouble of unloading is saved. The following simple diagram, of a single acre, shows the simplicity and perfection of this mode of operation at a moment's view. Any overseer can understand it in a minute's time, and a negro of ordinary intelligence is enabled to do the work, without any difficulty or inconvenience.

EXPLANATION.—The stars represent the heaps of manure, each containing ten bushels, placed in the centre of squares of 900 square superficial feet—giving forty-nine to each acre.

Thus it is seen with what perfect regularity and uniformity

the manure is hauled on the land. This done, we proceed to spread it out over the land, by first running off the rows with a scooter-plough in the old water furrow, which is yet perfectly visible, though the land lay last year in fallow—then two hands are put to each heap row of manure, with good shovels (Ames' long handles are best), and they scatter each heap for fifteen feet on all sides, which gives ten bushels of good manure to the surface of 900 square feet. All this is plain, simple, efficacious and practical; thus the broadcasting continues, until one suit of rows is done, when the ploughs commence, by first running around these rows with a scooter good and deep, and the balance is broken and bedded out with good turning-ploughs, by running four times in each row, thus dividing the soil equally, and throwing up each row uniformly. You thus see that the manure is incorporated equally and uniformly throughout all the soil. Whatever may be the opinion of casuists to the contrary, this is the true economy in the application of compost manures. I have given you, in this detail, the plan of operations that I pursue, in the preparation and application to the soil of compost manure; I shall, therefore, close this article with a few brief remarks upon the application and value of *guano.*

Ten to twelve years ago, guano, as a fertilizer, was comparatively unknown to the planters of the South. At the present time, however, the readers of the *Cotton Planter* are well "posted up" on the subject of its history and constituent elements, as a fertilizer. My attention was first directed to it in 1842, by an article from the pen of Prof. Liebig, prepared for the British Association of Agriculture. I was confident, after examining his analysis of guano, that it contained the true elements for the food of the cereals, and for the cotton plant, out of which to *perfect the seed.* I had it introduced into Alabama immediately, for experiment on cotton, the result of which experiment proved that I was not mistaken. The great secret of manuring cotton, like wheat, consists in making seed—the true object of the cotton planter should be, to make good, full and perfectly-matured cotton seed, as they (the seed), produce the cotton wool. On this interesting subject, we shall have more to say hereafter. Guano is a truly valuable fertilizer for grain and cotton, but to possess and use this valuable agency, we have to pay out a high price in gold, already made from our farms. It is a foreigner, and the cost of using it high. It can be used to great profit. Ten dollars worth of it, or 300 lbs. per acre, properly applied to land that will produce 500 to 1,000 lbs. of cotton per acre, will increase the crop from 1,500 to 2,500 lbs., due allowance being made for the casualties and vicissitudes affecting the cotton plant, as guano is no specific against any of the ills, but the lice and sore shin. Compost manure, prepared and applied as I do, and have herein described, produces the same results. The use of guano, then, becomes a question of policy altogether—just the same as whether it be the better policy of the cotton planter to purchase his bacon, for his operatives, from the West, or to raise it at home. Both are equally good when gotten into the meat house.

I have thoroughly tested guano, for the last ten or twelve years, on every variety of crops that we cultivate at the South. Its analysis sustains this position, had we no experience in its use. The best mode of application that I have found for using it is, first, to pulverize it, then add to it gypsum (sulphate of lime), in the proportion of one lb. gypsum to two lbs. of guano. For small grain, 200 lbs. of such compost, harrowed in with the grain, after thoroughly ploughing the land, produces a good crop. A heavier application will greatly improve the crop. For corn, 250 to 300 lbs., drilled along in the row, and then two furrows listed on it, and when you get ready to plant, open the ridge with a scooter and drop the corn, and cover as you desire. Thirty to forty bushels will be the produce, per acre, on land that, without the guano, might produce ten to fifteen bushels. For cotton, I have found it best to apply it in this way: first run off the rows, and then ridge with two scooter-furrows, by running round the row; upon this ridge scatter 300 to 400 lbs. of the compound, guano and gypsum, and then bed out the rows with turn-ploughs; then, when ready, plant your seed. Much of the success of using guano depends upon applying it early in the season, that it may become incorporated with the soil previous to the growing season. It may be applied, equally successful, without the gypsum—the gypsum, however, being cheap, can be used to advantage with it, as its application is, perhaps, always valuable.

DR. CLOUD.

CHAPTER III.

NATURAL HISTORY OF COTTON—ITS SPECIES AND VARIETIES.

SECTION I.—THE DIFFERENT SPECIES OF COTTON.

In dividing the genus Gossypium into species, we would follow Dr. F. B. Hamilton, (Linn. Trans. v, 8,) who says that the pubescence of the seeds is a better criterion than either the number and forms of the lobes of the leaf, or the number of glands for distinguishing the varieties. M. Rohn divides the cotton plants with which he was acquainted—

1. Into those with seeds black and rough.
2. Those with seeds brownish-black and veined.
3. Those with seeds sprinkled with short hairs.
4. Those with seeds completely covered with a close down.

According to Dr. Royle, who has been long engaged in the investigation of the subject in Great Britain and in India, the different varieties of the cotton may be classed under four distinct species, in the following manner :

1. *Gossypium indicum*, or *herbaceum*—the cotton plant of China, India, Arabia, Persia, Asia Minor, and some parts of Africa.
2. *Gossypium arboreum*—a tree-cotton indigenous to India.
3. *Gossypium barbadense*—the Mexican or West Indian

cotton; of which the Sea Island, New Orleans and Upland Georgia are varieties. It was long since introduced into the island of Bourbon, and thence into India; hence it acquired the name of Bourbon cotton.

4. *Gossypium Peruvianum, or accuminatum*—which yields the Pernambuco, Peruvian, Maranham, and Brazilian cotton; especially distinguished by its black seeds, which adhere firmly together. This variety has long since been introduced into India.

The chief varieties cultivated in the United States, are the black seed or Sea Island, (*G. arboreum,*) known also by the name of "long staple," from its fine, white, silky appearance and long fibres; the green seed, (*G. herbaceum,*) called the "short staple," from its shorter white staple, with green seeds, and commercially known by the name of Upland cotton; and two kinds of Nankin or Yellow, (*G. barbadense,*) the Mexican and Petit Gulf.—(*Pat. Off. Rep. for* '53, *Ag. Dep.* 179.)

SECTION II.—THE COTTON PLANT—"SEA ISLAND" COTTON.

From the Southern Cultivator.

THERE are various statements in regard to the number of *species* of the cotton plant. Some authors assert that there are not more than eight, while others affirm that there are upwards of a hundred; and indeed that there is no end to them. It is not at all likely that anything certain is known about the matter, botanists never having taken the trouble to cultivate a great variety of them in order to ascertain the difference between the several kinds. We believe, however, that attempts have been made to do this to some extent in Jamaica, Trinidad and St. Vincent's, where the various plants of the numerous regions where the cotton shrub is found, have been grown side by side in private gardens, on plantations, and in botanical collections. Of these, the garden in Trinidad alone remains, and we are not aware that any great advantage has been derived from that. This much, however, we know, that the several species differ materially in appearance, varying from four or five, to fifteen or sixteen feet high.

There is no doubt that the plant was well known in ancient times; but at what period it was introduced into America, we are not precisely informed. The Sea Island cotton is the produce of a plant that seems to have been first carried to the Bahamas from the island of Anquilla, (whither it is believed to have been transported from Persia,) and was sent to Georgia in 1786. But there is evidence of the existence of the cotton plant in America long before there was any direct communication between the civilized world and the two great portions of this continent; and it is a well-authenticated fact that the Spaniards found cotton cloth, or calico, a common article of dress among the inhabitants of Mexico, upon their first invasion of that country. *Calico* obtained its name from *Calicat*—an insignificant town in India, where it was probably first made. It was an article too expensive to be purchased by the laboring classes, on its first introduction into England; and it was little imagined, in the early days of its manufacture, how wonderfully it was destined to alter the whole face of commerce and society, and become the great staple commodity of the western hemisphere.

In China, it does not appear to have been employed to constitute articles of dress before the thirteenth century. In Spain, it is believed that the Moors employed the filaments of cotton for weaving cloth in the tenth century; but the quarrels between the Mohammedans and the Christians kept the rest of Europe in ignorance of its manufacture for many ages. Italy was the first to adopt it, and when the genius of Arkwright, Hargreaves and Cartwright had invented the proper machinery, England turned her attention to it in earnest, and gave an impetus to the growth and manufacture of cotton, which has since been constantly on the increase.

The cotton used for manufacturing purposes is distinguished by the length and shortness, the silkiness and coarseness, and the strength and weakness of its several filaments. Some species of the plant thrive best where they can have the benefit of the sea air, and the produce is fine in proportion to their nearness or distance from the coast. Others, again, require the interior of the country. In dry climates, as on the mountain-bound shores of Brazil, the best plants are met with on the coast; while in damp climates, like that of Pernambuco, the most valuable cotton is obtained from the interior.

But, whether seen bordering the lofty acclivities of the Andes, with the wide Pacific heaving its boundless waves to a limitless horizon, beneath a sky of more than Italian azure, or met with in the broad, rich valleys and on the sunny uplands of our own beloved country, a field of cotton in full bloom, with its dark green leaves and snowy pods, (with here and there a magnificent magnolia or a noble pine, rearing its lofty head into the air), is a beautiful sight, more especially in the "picking season," when hosts of busy hands are gathering the valuable produce, and preparing it to enrich and comfort the inhabitants of our own and far distant lands.

The finest and best kind, is grown on the low sandy islands off the coasts of Georgia, Florida and South Carolina. It is sometimes called "*black* seed cotton," from the seed contained in the pods being black; while the seeds of the staple cotton is called the "*green* seed cotton," for a similar reason. The "staple," or filament of Sea Island cotton is exceedingly long, silken and delicate; and it commands a high price in market—

more than double that of the short staple. Great improvements have been made in the latter by assiduous and careful cultivators, by the selection of seed; and its culture on fair lands not exhausted by a ruinous system of tillage, is believed to be far more profitable than the culture of Sea Island, owing to its superior yield and the facility with which it may be ginned and prepared for market. D. R.

SECTION III.—COTTON SEED.

Mr. Editor:—I am fully aware that many of my agricultural brethren will think I am actuated by sordid motives, while others will think I am fostering what is truly humbuggery,—sales of cotton seed or anything else at exorbitant rates. Should we all be so fearful of censure, as to advance nothing but what is received as known by the vast body of agriculturists, we would always travel in a circle. I will, therefore, venture to give my opinions on the subject of cotton seed.

To be able to produce for sale two millions of bales, we must cultivate full 2,500,000 acres to cotton. Suppose we could, by dint of improved seed, increase the per acre yield, or the per hundred turn-out of lint only a few cents per acre, the gain would be immense. But suppose we could increase every one-hundred acre planter's yield 100 lbs. of cotton per acre, we would add to his income a clear gain of $100 to $200. Is this possible? I answer, afte rtrying to improve seed for about fifteen years, that I believe we can.

Improved cotton seed attracted public attention in this country, near on to thirty years ago; and I well bear in mind when Hollingshed cotton seed sold in Carolina. They were nothing else but Mexican seeds; and nearly all the improved seeds are of that kind. Why planters will hesitate to use the improved seeds, when they use improved cattle, horses, hogs, sheep, poultry, fruit, &c., &c., is more than I can understand. Will any one hesitate to admit that there is a vast gain by planting one kind of corn over another? Yet the improvement in cotton seed is fully equal. That humbugging has been practised in cotton seed, as well as in sheep, *morus multicaulis*, hogs, &c., I know, and yet I believe good has been done. Large prices induce attention to be directed to the production of choice seed. Many persons rate cotton seed, as manure, and as food, to be worth sixteen cents per bushel. I have known ninety-eight bushels of corn grown with, say 250 bushels of cotton seed, there being full one-half the value left in the soil, where not over fifty could, or ever did grow without the seed; this is rating cotton seed at twenty cents or more. Who would go to the trouble of making seed for sale at a profit over this of a few cents when thereby he is injuring his land? It may be said, all can improve? Granted. But can all improve seed as cheap as they can buy. I do not believe they can, because the man who can sell $500 or $1000 worth, can bear extra labor and expense. I know that extra labor and expense, of course, must be borne to make seeds that are the best to plant. If seeds are not thoroughly cured, they are injured by transportation, by being in a large pile. All of my seeds are sunned after ginning, and a hand does nothing else but attend to them. Even if a hand can stir up and attend to 500 bushels per day, there is scaffolding, and house-room, and attention, and no man is willing to do this for nothing. A bushel will plant two acres of land. I have planted this year forty to sixty acres thus, and plant every year, more or less acres thus—even at $1 per bushel, the seed only costs fifty cents per acre, and if the gain be only ten pounds of cotton, the planter makes a great

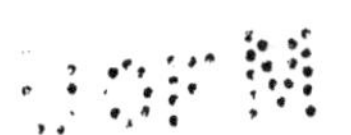

per cent., having the next year's seed to plant other acres, say, at least twenty acres for every original acre.

The seed mostly relied on in Mississippi and Louisiana, are Mexican seed, known in Carolina and Georgia as Petit Gulf seed, because there planted and improved. In the hills around Rodney, Miss., the improvement began, and there are just as good seed at present elsewhere, as there is now near Rodney. We plant Sugar Loaf, or Prolific, Lewis' Prolific, Vicks' 100-seed, Guatemala, a seed not of Mexican origin, Brown seed, and others. Except the Guatemala, they are all, I believe, mere selections from the Mexican.

I do not pretend to affirm that any of these seeds will produce, in quality or quantity, so much greater than seed usually cultivated in the interior of Mississippi, in Alabama, Georgia, or Carolina, as to warrant the planter in giving such rates as \$3 or \$5 per bushel! The grower of the seed deserves a portion of the increased value, but the planter (purchaser), also deserves a fair portion, and the greater. I know from repeated trials, that good seed will produce, say as six to seven, that is, an acre which would produce 600 lbs., with ordinary. seed, has produced here 700 lbs. And I know of planters who will not buy seeds, yet will haul them twenty to thirty miles, if given to them, clearly showing their real opinion of the advantage derivable. I believe the yield is greater over common seed. I sincerely believe I have laid out 75 cents for a bushel of seed, and made more thereby than any other investment; and I think by buying seed to plant, say one-fifth of one's crop yearly, that any planter will make thereby; I mean where seeds are used for ten years or so, of any one kind, of the ordinary kinds. A planter who plants 100 acres, may buy seed for twenty acres, say ten bushels, at \$1 per year, and make a better investment than buying a negro fellow at \$500. Some high-strung planters believe it

is a departure from ancient custom to sell seed; so it may be, but if there is a real gain to the public, the man who adds to that gain, is a public benefactor, whether he is reported to be a regular planter, or a mere huckster of seed. I have given away more of these gains to the public than any high set planter in the South; and as I make neither credit nor glory, nor cash, from such drones, I am very content to receive their obloquy. I may err, but it takes two people to be in error, ere I can inflict an evil; and whilst I think my country can be a gainer by sales of seeds or chips, of anything, I will urge the matter, and only ask a trial.

Edwards, Miss., April, 1848. M. W. PHILIPS.

SECTION IV.—SUGAR LOAF COTTON.

MR. EDITOR:—I beg to call the attention of the planting interest to the early maturity and the productiveness of the cotton called in Mississippi, Sugar Loaf and Prolific. I do so at this present time, that all may be able to obtain reliable facts, in season for the next planting. From my present knowledge, I do not hesitate to recommend it warmly for the above valuable considerations. I have planted it only in 1847 and '48, and have no more personal experience than the yield of one crop, and the present prospect. I do not promise for it the great yield that the seller of the Mastodon seed did for that seed, nor the yield that was promised from the Turin and Okra cotton; but I do say, that I believe it will pay the planter, even if his seeding costs him $1 per acre. More than this, I leave others to say.

This day, being called into my field, south of my pasture, and some two hundred and fifty yards from another field in cultivation, where I have my selected Sugar Loaf seed planted,

I was so forcibly struck with the prospect, that I conceive it my duty to draw attention thereto. I saw repeatedly limbs, with six, eight, ten and twelve bolls and forms, which were not that many inches long, I could span so as to touch ten without any exertion. I have forty acres planted to the Sugar Loaf seed, and think I reasonably calculate, from present appearance, on fifty bales, and I don't think any other forty acres of Petit Gulf seed, promises forty bales. My seed have been planted remote from others for these two years, they were selected from the field by myself and an old negro woman; yet, I find a great tendency to run back, and which can only be guarded against by careful yearly selection.

I have many friends who are planting it, and they pronounce two weeks earlier in maturity, a great gain when the army-worm is expected.

The picking qualities. I can pick 200 lbs. per day, easier than from ordinary Mexican. (We term the improved cotton Mexican, which is known in Carolina and Georgia as Petit Gulf, because everybody who sells seed, marks his bags Petit Gulf, the first improved seed emanating from that section.) I can gather 150 lbs. This was the fact last fall.

Those who are skeptical will consult their own interest by writing to friends in Mississippi, where the seeds are well known. There are a great many here who are now planting the seed, and as they do not sell seed, and are too proud to advertise, no doubt but that their evidence will be good.

The seeds I have, were presented to me by Mr. Farmer, living in Yallobusha County, I think. He was the first in this country to call attention to the seed some four years ago. They are no discovery of mine, nor have I improved them. I only claim calling your attention to what I believe will benefit you, sir.

I believe the seed that Dr. Cloud plants to be no other than this same variety of the Mexican; and if so, by carefully selecting and keeping as pure as possible, he will do as much good by selling them at $1 per bushel, as he has done by calling attention to manuring, and I believe he has done more good to many sections of the cotton-growing region, than all other writers together. It is necessary that one should be very zealous in a cause to get attention, and I hope all growlers will confine themselves to the growl, while we doctors, (humbugs, if one of the savans desire the name,) are allowed to keep in sight of facts, even if at a distance.

I am sincerely a friend to the cause of improved agriculture.

Edwards, Miss., July, 1848. M. W. PHILIPS.

SECTION V.—IMPROVED COTTON SEED.

MR. EDITOR:—I have received and answered about twenty letters in relation to cotton seed; and as those letters are entirely from your subscribers—and, by-the-by, the postage always paid, too—I beg a corner in your right worthy paper, that I may save all trouble by those who desire cotton seed.

I will add some remarks about the various improved seeds grown in this section, and make it my endeavor to fairly state my opinion of them. I beg, also, to say, that I will very cheerfully purchase any kind of seed desired by your subscribers, charging only what I give; and should it be necessary to incur any expense in selecting, I will divide expenses between the parties ordering. I mean by this not to force any one to take my seed, unless they are satisfied, and yet am anxious that all should have the benefit of improved seed. I will pledge my good pen, to show in this country as great an improvement in the quantity or in the quality, and in the picking qualities, as could be reasonably expected; but, in some instances, the price of seed is too high. My personal friends have the seed for sale. I do not disparage the seed, but I deem one dollar per bushel, at the gin, to be a remunerating price, and at that any planter can afford to buy a few. I would advise no one to buy many until he has proved them. I have labored too long and too earnestly, to help my native South, to now risk all for the purpose of making a few dimes. I do not offer my seeds as being the best, nor as yielding two and three bales per acre. I only pronounce them as being very carefully sunned, both before and after ginning, and to be seed from the best I could procure. They yield well here; I say not what they may do in Carolina or Georgia. I could name friends in both States who have planted my seed, and who write me very favorable accounts. I do not desire any puffing of my wares, lest I may be again blamed. And in this matter I claim to be useful, in offering an article that is needed, and, if I am not in error, an article that will benefit the purchaser more than the seller.

I charge, in all instances, to home folks or to strangers, as follows: Petit Gulf, or Mexican, generally known here as the latter, fifty cents; Sugar Loaf, from seed carefully selected under my own eye, $1; Vick's 100-seed, $1.

When I send off, I put in five bushel sacks, made of Lowell goods, costing me twelve and a-half cents, the sacks to weigh 125 lbs. or over, if I wish, and charge fifty cents per sack, for the sack and hauling to Edwards' depot. Col. Vick charges $2. I did not learn this until I had fixed my rates; and I have too few to affect him, who kindly presented me with my start.

The Petit Gulf seeds are only Mexican seed, acclimated and selected. When these seeds obtained their first celebrity, it was usual to select seeds from the pile, for their white color and small size. Latterly, we pay more attention to the production, quality of lint, and picking qualities.

To Col. Henry W. Vick is, therefore, due the credit of first, scientifically, with great personal labor, perseverance and skill, making the proper selection. We regard these as a very decided improvement, and his selling at $10, for two years, proves such to be the fact. I have planted them two years, and will plant one-half my next crop with them.

The Sugar Loaf was introduced by Mr. W. B. Farmer, Last Chance, Miss., from whence I know not. Two years ago, he kindly gave me two bushels. I planted them on four acres—the product, early maturity and extraordinary picking qualities, pleased me so much that I took my most careful hand, and together we selected enough to plant near twenty acres. It is this seed I offer. There are many planters who put the gain at fifty, seventy-five, and even one hundred per cent. I do not promise that much. Having fairly tested it, I place it, say two to four cwt. more per acre on rich land, and enough to warrant a trial.

The Brown seed is said, by its friends, to excel the Sugar Loaf in all its good qualities. It originated in Copiah County, I learn.

The Tarver seed, from Alabama, is greatly praised there. These three latter seeds I planted side by side, and I will hold to the first; and, as I have no prejudice to uphold, I presume I am correct in my judgment. How these latter sell I know not. I learn that Sugar Loaf sold at $1.50 last year.

Hogan seeds were introduced into Mississippi by Mr. Wm. Hogan, who lives a few miles from me. I saw his field, of some fifteen acres, and was so well pleased that I purchased one bushel for myself, and two and a-half for my friends, at $10 per bushel of twenty-five lbs. As a special favor, I was allowed the seed cotton, selected from special stalks, and for the purpose of testing the yield of lint. September picking gave, viz.: 116 lbs. yielded thirty-six lbs. of lint, or a trifle over thirty-one per cent., the largest yield I ever found, and I have been thus testing all cotton for many years. His price is $10.

Banana cotton, introduced by my friend, Col. Hebrun, of Warren County. I saw his cotton and his book. I also saw a piece of the same, owned by Dr. E. Bryan, his neighbor. Two other gentlemen, with the above, Mr. Cook and Mr. Gibson, being the only growers at present. Of these two last kinds I know the history, but am not at liberty to say more, than that the production exceeds anything I have ever seen. The Banana seeds are held at $100 per bushel, and no less than a peck can be sold by written agreement.

Pitt's Prolific I never saw, but am told by its friends, that that seed will be planted, and a wager put up, that the product shall exceed any of the above. The price, and where to be obtained, I am ignorant of, but in another year I will have tested, and can report.

I saw three acres of the Hogan seed at Judge Pearce Noland's, a large planter near me, and the planter who gave to the Petit Gulf its deserved notoriety. The Judge's account of what he picked, warrants me to say to any one who regards any improvement in cotton seed as a humbug, that he can be well paid for his year's labor and time, if he will send a few dimes out, and superintend the culture and picking; that is, if he is right. In other words, I think he can get a very snug crop for doing nothing, but to see there is no cheating him. But let me caution him, lest he may lose a crop; for some of this seed will certainly produce two to five, or may be ten cwt. more per acre, side by side, than the best Mexican yet grown and not improved by selection—or than any other seed not here named.

I hope the above may do some good, if no other but to incite your readers to a closer attention in improving their seed. My remarks are open for examination. I give, to the best of my ability, what I think is true; and, as I have not $500 worth of seed saved here for sale, and not $1000 for the two past years, I earnestly beg that the liberal of my profession—agriculture—will not attach to me any desire to humbug; $1200 is too small a bait—and I hope, by giving satisfaction with these few seeds, to open a yearly mart for $500. Therefore, I would not kill or injure the goose—the rather I would feed her high—with good weights and sound seed.

Excuse the great length, but, if I am right, the public advantage will be a sufficient excuse for

Edwards, Miss., Nov., 1848. M. W. PHILIPS.

SECTION VI.—IMPROVED COTTON SEED.

MR. EDITOR:—Having already received inquiries concerning cotton seed, I beg again that you will favor me with inserting my reply. It will save your subscribers some writing, and myself a great deal. I have had more or less to do with a printing-office for several years, and I have had advertisements slided on me as communications. I do not do this; I send this forth as an advertisement. It will aid me greatly, and will aid your subscribers a little—probably as much as myself.

I will be prepared to fill orders to a limited extent, but will not reserve as much seed as I did last year, unless orders come in before Nov. 1. I will record orders as received. All seed sent from here shall be thoroughly sunned before and after ginning, and well cured before being put into bulk. My charge will be invariably $1 per bushel of 25 lbs., 10 cents per bushel for the sacks and hauling to depot. No seed will be reserved except from cotton picked in September and October.

Now as to varieties:

The *Sugar Loaf Seed* will be from the second year's selection. Sugar Loaf is regarded, by every neighbor I have, as being the best seed yet planted by them. I have heard the opinion of a majority of them. With me they prove, on rich land, the very best.

Vick's 100-*Seed* is generally acknowledged to be the best selection from the Mexican or Petit Gulf, ever planted in Mississippi. Col. Vick sold his, the present year, at $1.50 per 50 bushels, or over, and $2 for any quantity under.

Brown's Seed is, in my opinion, identical with the *Tarver Seed* of Alabama, and very much like Sugar Loaf; bolls more pointed; not so prolific or so easy to pick.

Pitt's Prolific, I have growing this year, for the first. For me it does not well, and does not seem to be established; some stalks are good, some excellent, others so-so.

Hogan's.—I have eight acres to itself, no other seed planted; besides two other patches replanted with other seed. I shall only reserve from the first. I am not so highly pleased as last year, but there is great allowance to be made: I had to replant twice, not cleaned out until May, and the land was in part too level—a good part of the land so level that cotton was almost drained out.

Banana, from seeing and feeling, I pronounce identical to the above.

Prout, the same, I might say, as I can show both kinds in same field.

Chester, I might say as for Banana.

Pomegranate, I believe to be identical: never heard of it until I saw a notice from a Mobile paper, though Gen. Mitchell lives not far from me; and never heard of it since, except through distant papers.

For the information of some, *Banana was not planted in Warren County, Miss., before* 1848, and was then planted by four gentlemen—Messrs. Hebrun, Bryant, Cook, and Gibson.

Multibolus.—I have some 100 stalks, and some of them exceed anything I have seen. The introducer promises the yield of lint to be some forty per cent. It is good certainly thus far.

Rob Smith's 25-cents is very prolific indeed, with long tortuous limbs, leaves silky, bolls slim. I have but little.

Mammoth—Also from my friend Smith: very large bolls, and quite prolific.

Any farther information will be cheerfully given.

I advise early application, as I am determined to sell only the best, and not to reserve many. Orders must be accompanied with the cash, or some certain means. Payment first of January will be time enough, but it must be certain.

To merchants ordering say 500 bushels, I will deliver at depot for $1, so that planters can buy at my price, and yet ten per cent. be realized.

I shall plant nothing but selected seed; thus planters may reasonably expect the purest seed, according to my judgment.

With great respect, I am yours, &c.,

Edwards, Miss., August, 1849. M. W. PHILIPS.

SECTION VII.—BANANA COTTON SEED.

From the Southern Cultivator.

MR. EDITOR:—My remarks as to the above variety of cotton seed, as published in the *Cultivator* for last November, have been deemed by some persons of my acquaintance, as having emanated from a desire to put down other seed, that I might sell. To such, I have no reply.

Duty to the growers of said seed, requires of me to say what was the fact, and to place the error where due. Understand me. Col. John Hebrun and Mr. David Gibson, of Warren County, Miss., are personally on the very best terms—they are the principal growers. There is no issue between us, as they know my motives and the facts.

The Banana seed of October, 1848, I saw. I culled a few seeds, and planted in 1849. I pronounced them identical with Hogan, and *they were.* Mr. David Gibson had procured a variety of seed—not from Mississippi, Louisiana, or Alabama—and in his judgment they proved, planted side by side, to be superior to the then Banana; and knowing the sale of the said seed would be ended with the one year's growth, he refused to join in sales, but proposed to supply his new seed. Thus was a cotton which I never saw until ten days or so ago, sold as *the* Banana. Was I to blame? I did deem the offering of a seed by a new name, for a large price, a wrong; and I deemed it my duty to expose. I did so, and will do it again.

As to the present Banana, I saw the field from which forty-five bales were gathered. Mr. David Gibson is practically conversant with surveyor's implements, he is a correct man, he assures me there are thirty-eight and a-half acres, *measured.* The growth and appearance is very similar to the Hogan, and I doubt not, had the same primary parentage, but the Hogan was taken to Alabama, thence to Warren—the Banana, directly to the hills of Warren, eminently inducive to short joints, yield and quality of lint. Mr. Gibson will procure for me a daguerreotype of two branches I saw at his house, which will be sent to you to copy in your Journal. It will be done at my request, supposing it would be ornamental, and serve to show how the cotton can grow. Letters written to David Gibson, will be attended to. I trust this act of justice will not be misunderstood. I am sure the seed will be an acquisition, and as to price, every man has the right to charge what he pleases—puschasers to pay, or not.

Yours, with respect,

Edwards, Miss., Jan. 1850. M. W. PHILIPS.

SECTION VIII.—SILK COTTON.

From the Southern Cultivator.

MR. EDITOR:—Your last Number is mislaid by some one, though in the house. I therefore cannot refer to the page on which my friend, J. V. J., in alluding to the seed I sent him, gives me credit for more than I deserve. I beg to allude thereto. First—I object to the name Early Sugar Loaf. I detest multiplying names. I received the seed from Mr. Farmer, Hard Times, Miss., as a present. I have selected from the field for four years. Those I sent J. V. J. were of the third year's selection, and such seed as I never sell, except in very small parcels, and then to my favored few: because I cannot be paid for doing it, and I only select some ten or fifteen bushels yearly—with these I plant twenty to thirty acres—from those I select the ensuing year, and *from no others.* I still call mine Sugar Loaf, and sometimes Select Sugar Loaf.

The Silk Cotton seed sent to J. V. J. was grown here, the second crop from seed sent me by my friend Col. H. W. Vick, of Vicksburg. *It is not the* Silk Cotton *of the south-west.* Let it be understood. J. V. J. has probably the only seed now in existence. Col. Vick selected the seed from his 100-seed, and from the Mexican, for several years. He projected with the seed, he said, resembling silk in feel, and quit, I think. I selected one year, and the second selected for J. V. J., and had so much to think of, I cast all off. Therefore, Col. Vick has the credit, and I must not take it to myself. Col. V. deserves more credit as an observing planter, than any man who has dabbled in experiments. He has been improving seed these fifteen years. And proof—he works fifty hands, or over, and made, the past year, as bad as it was, nine bales per hand! Tell that to the B'hoys who think there is no virtue in improving seed.

I have required J. V. J. to send me a bushel of the seed. This I will project with, and return him the progeny, even bettered; for I assure you and all others, that, in Warren County, and a part of Hinds, we can improve any seed. Our climate and soil, and our attention, is something.

I will suggest to all parties interested, that we name the seed, (as the cotton is, beyond doubt, distinct,) Jethro Seed—after JETHRO, of the *Cultivator*, whose anonymous contributions to that periodical are considered among its best.

I ask it of friend J. V. J., and will send him a letter of introduction to Jethro; and, by-the-by, I hope there are many others who admire the writings of Jethro, and who miss him as much as does your friend.

And, by-the-by, as I have returned to my duty, I would like to see my old companions-in-arms—my noble *Coke*, caustic *Broomsedge*, and others. Come, friends! let us, one and all, give a whole-soul life to the *Cultivator*, this year. Our readers are many. They are our brothers, of "our own, our native land." We are aiding the readers and ourselves. I am the oldest man among ye, and I hope you will not leave the load to the "old man"—my nickname, when about twelve years old, and followed me up. May our cause prosper—may our whole country live together a thousand years, and peace be with ye all! Sincerely yours,

March, 1850. M. W. PHILIPS.

SECTION IX.—MULTIFLORA COTTON—"MONEY BUSH."

MR. EDITOR :—The March number of your paper is just at hand, and I must needs thank you for the high estimate you place upon me. I can assure you, I will try to deserve your encomiums, and to meet the expectations of my brethren.

This thing I say—and do not think there is one man from Maine to Texas will think I say falsely—I love my calling and the progress of my brethren too well, to designedly lead them astray. I have no unkind feeling for man or woman, that should lead me to molest or do injury; nor have I any jealousy of other men of high standing, that I would wish to lower them that I might be seen.

I am induced to make these remarks, because I will fearlessly expose any subterfuge that may be resorted to in the way of humbugs or deceptions. I have learnt that some rather strong remarks are made against me in the January number, which I have not seen. I will see it as soon as I can procure a number, and will reply. I fear no one, for I am determined to be actuated by a higher motive than mere words, or making dimes.

There is yet another seed called Multiflora, which, I am informed by one of the growers, was procured from Mr. Prout, of Alabama, and is, therefore, the same seed as the Hogan, Cluster, &c., &c. Why not say, openly and candidly, that the seeds are Hogan, or Prout, or Cluster, and then ask their $2½ or $5 per bushel?—thereby, any man buying would understand what they were getting. I have the word of a friend who planted Money Bush, that it was also the same cotton.

I put the question to any man—how would you like to buy, at one, or two, or three, or five dollars per bushel, the same seed, from two or three *gentlemen*, under different names, each one declaring their cotton to be "the best in the world?" You would not like it, of course. Then act squarely to all men; call your seed right, and ask any price you please.

M. W. PHILIPS.

SECTION X.—VARIETIES OF COTTON SEED.

MR. EDITOR:—Last spring I received a lot, say about one hundred cotton seed, from G. W. Mabry, of Vernon, Mississippi, called Multiflora. I planted them myself, and gave them two workings. The product was so good that I saved all the seed. The growth I conceived similar in all respects to the Cluster or Hogan, &c., &c., yet the seed were all white. I learnt they were procured from Carroll. This season I saw a gentleman from thence, who had seed to sell. I inquired as to the history, and traced them through the same Carroll County source, to H. W. Prout, and have therefore pronounced them identical with the Cluster. My friend, G. W. M., says I am in error, as these seeds remain white, but the Cluster will in three or four years run into green. I state this, that all sides may have a hearing. I have received this season as a present:

Royal Cluster, from R. W. Harris, Greensboro', Alabama.

Golden Chaff, from G. W. Summerville, Hope, Alabama.

Multiflora, from W. W. Whitehead, Middleton, Miss.

Seed, second year from Mexico, from Jas. E. Harrison, Aberdeen, Miss.

Seed from Texas, I think, from "Hinds," Cayuga, Miss.

Brown, from H. W. Griffith, Hinds County, Miss.

Willow, from W. Montgomery, Hinds County, Miss.

Selected Seed, from Col. Jno. L. Croom, Greensboro', Ala.

Cotton Seed from China, from Patent Office, Washington City.

Guinea Seed, from John A. Heard, Hinds County, Miss.

Magnolia, from A. N. Mayer, Holley Springs, Miss.

These seeds are each generally highly recommended, and some of them are spoken of in such terms that I am induced to expect great things of them. I will thus, with others in field culture, have some twelve or fifteen varieties—at least with different names—and will be able to report in the fall, I have no idea that all my correspondents can be pleased with my report for it is impossible for me to suit each kind to land, unless my friends had given me hints to guide me in the selection of spots to suit each. But I hope they will bear in mind, that I can have no interest, other than in selecting the very best, and that no one locality will suit every variety. I will illustrate by an example: Some of my friends declare that Sugar Loaf will "double almost" any other variety, while others declare it is no account; all agree in early maturity and ease of picking. The reason of these different *true* statements arises from the fact, that Sugar Loaf does best upon rich, fresh land, inclining to moisture, that is, green land; whereas upon high and dry land, and old at that, the production is not good.

J. E. H., asserts of his seed, that the production and yield of lint is greater than any I have tried; he states figures; he did not state quality of land; I have planted them upon good upland, cleared twenty-two years ago, from which was taken a 500 lb. bale a few years since. I only anticipate to give relative yield, and by testing two or three years I can find out which land suits each kind.

The adapting of seed to the soil that suits is no small matter, and does not receive the attention it deserves. It is impossible that any one man can experiment so as to arrive at just conclusions. He may take one, or two varieties, and test; but to test the great many varieties that we now have requires too much time and trouble. If some of the agricultural societies would take the matter in hand, trouble would be divided, and just conclusions arrived at. I have bestowed considerable attention to these matters, and have interfered with business matters to do so. I will continue to prosecute my experiments, deeming the encouragement I have received, from your subscribers especially, as claiming my time and attention. I would earnestly request of all persons who have new seed, to note particularly the product upon different soils, and give me their views this fall. Some one or two of my friends have complained of the 100-seed; why this is so, I cannot understand—100-seed is certainly the best variety of the Mexican, and is recommended as such. Col. Vick took the best Mexican—the little hum Mexican—and from that variety after several years' selection settled down upon the variety he named 100-seed.

This variety, so far as I have experienced, is the best upon rich or poor land—compared with Mexican or Petit Gulf. Unless Brown seed excels 100-seed upon rich upland, I would prefer that variety for that land, or even upon rich dry low ground. The Sugar Loaf will do better upon rich fresh land, but after a few years' culture the Vick seed will excel it. And this is Col. Vick's opinion, I think.

Bear with my many words; I merely intended to have thanked those friends who so kindly favored me with their select specimens, and have thus ran on. I beg they will accept my thanks, as also all others who have sent me other articles upon trial. I will endeavor to give a just and true account of them.

I have near about twenty varieties by name, and hope to be able to make an interesting report next winter.

Since writing the above, some good fellow has sent me seed of a millet, entirely new to me. I have seen and grown several varieties, but the shape of seed, resembling slightly wheat when not grown, or injured, or perhaps rye when cut too early—the seed being longest and largest at one end. The seed came from Lauderdale, Mississippi, and the writer gave "Lauderdale" as his signature. This mode of signatures does in some countries; for instance the Duke of Devonshire, might sign his name Devonshire and be known. But, *de gustibus non*, I will not quarrel with any kind-hearted fellow who will thus make me a recipient of his favor.

Truly yours,

Edwards, Miss., April, 1850. M. W. PHILIPS.

SECTION XI.—SCRAPER AND COTTON SEED.

MR. EDITOR:—To-night I read your May number, in which I notice a call for the description of the Mississippi Scraper, or a drawing thereof. I also received a letter to-night, from a new friend in Alabama, on the same subject.

I hasten now to say to you, that I will write to Vicksburg to-night, and endeavor to get a drawing. If I fail, I will send a scraper to Augusta, Georgia, with direction that a drawing be taken, and that the scraper be presented, in my name, to the contributor of the *Cultivator,* who is most punctual in supplying original matter for this paper.

I have used the scraper for ten years, and believe I had

the first one ever tried in Mississippi. I think I tried one in 1838 or 1839, and continued trying to improve it, until Smith Taylor, a blacksmith of Jackson, Mississippi, beat me so bad, I had to take his. The suggestion was first made to my brother, in October or November, 1837, by a Mr. Tilgman, of Tennessee, who had to take log-cabin fare for a night. I deem it proper to say, that many planters here say they scrape with the turning-plough as well. I have tried both, and as one is as cheap as the other, and the scraper works best with me, I retain it.

I am positively certain that upon land put in first-rate condition for scraping, that I can have eighteen acres scraped in six days, by one hand. I mean he can average three acres per day, for six days. Of course the land must not be wet, and in the condition that we can have it nine years out of ten.

Allow me, *en passant*, to say to Hinds, of Cayuga, Miss., that he is not many miles from me, and if he will come this way, that I will show him one-fourth of my crop planted in Hogan seed. The Banana, Hogan, and Pitt—"identical!" "All in my eye and Betty Martin too!" The Pitt is as much unlike Hogan, as Sugar Loaf is unlike Mexican. They have not leaf, stalk, or boll alike. There has been too much speculation in seed; but my dear fellow, let us be cautious lest we do injury otherwise. You have heard of the Montgomery on Fourteen Mile Creek—the elder brother, A. K. Montgomery, says his Hogan excelled, last year, Mexican, nearly two-fold. His father, an aged planter, plants it this year, after his experience last year. Do you know H. W. Griffith, between Palestine and Utica? He will assure you that he exceeded a bale per acre, last year, with the Brown seed. David Gibson, of Warren, exceeded a bale per acre, of the Banana. And more I might name, but I suppose enough for the occasion.

The Banana is a seed introduced by David Gibson, of Warren County, Miss. I saw the correspondence, and assure Hinds that these seed are introduced from Georgia, yet I believe they are of the same parentage as the Cluster, which is the original name, and from whence Wm. Hogan, of Warren, procured his seed.

The Multiboll I never saw, and therefore cannot say aught for or against it. But the Sugar Loaf, upon rich fresh land, say big black, or Mississippi low grounds, will excel any other that has yet been tested by its side, and I know not a solitary person who denies it.

As to those, whom Hinds charges with "always collecting new corn and cotton seed," and they being the persons who fail in cotton crops. That may be so. There has to be one sheep in the flock to carry the bell, and provided the bell is useful, the other sheep should not complain. One who knows all this ought not to split upon the breakers.

Hinds has done me the kindness to send me some seed, which I planted, and have only two or three stalks. They shall be nursed, though I do fail in a crop. I am fond of these sort of things, and as it is necessary that there should be somebody fool enough to waste his time, I might as well be that one, as I have no babies to feed. So Mr. Hinds, thou neighbor of mine, who blaze away with a scattering shot gun, under so large a name, e'en load up and fire again—no telling, you will hit somebody; and it will keep up our blood in the long days now setting in.

I would like to know ye; you talk well about impositions. I will join you in some respects, but I prefer to be specific; and I will not, if possible, commit a similar blunder, as any one who says Banana, and Hogan, and Pitt, are one and the same. The first two may be, but never the last.

I have been buying seed for some fifteen to seventeen years. I have sold for two or three years, and I wish to sell again.

I do try all sorts that are recommended to me, and some that I select myself. Now, Hinds, here is a customer, and an old one, who can stand much shooting at.

With respect, yours, &c.,

May 24, 1850. M. W. PHILIPS.

SECTION XII.—THE DIFFERENT VARIETIES OF COTTON SEED.

DR. CLOUD :—Many ask me what is my opinion now of the different varieties of seed, and to save much writing, I ask of you the favor to be allowed to answer one and all.

There are those who ridicule selling improved seed, but they will plant such if given ; others ridicule, to be thought of the prudent sort of folk. Every one to his notion. In 1833, or about—for it was in '32 or '33—there were those who ridiculed my trying to get up a better seed ; they were only my second neighbors. This class has greatly increased, and even improving men lend themselves to this cant. My improving was only intended for home consumption, and would have so continued, had not an estimable friend, an old school-mate, insisted, if I desired to benefit planters, that I could do more by selling seed than any other way. He had tried seed grown here, perhaps two years, being sent to him as an old and cherished friend. Others inquiring, put me in the way of selling seed. Of this it matters not ; a planter might as well sell seed of corn, oats, &c., as cotton. It is all sheer ruffle-shirt cant, to ridicule selling anything a man has to spare. To cull seed carefully, cure them properly, attend to correspondence, and all the little perplexities, as well as loss to be incurred if a full crop is not made, is not very satisfactory to one, unless the almighty dollar has complete possession of him. At least, I am willing to quit the business.

Some three years ago, I offered a near neighbor, and a dear friend, all my improvement, if he would take the trouble off my hands; and I will do so to any planter who will assure me of his devotedness to this matter. The man who is governed only by cent. per cent. will not do. I plant this year near 200 acres, or perhaps over, of select seed. I think I make by it; selling seed is too small a business—yet to be called the "celebrated cotton-man of Edwards, Miss.," is enough to induce any one to persevere.

All this, by the way. Ridicule may turn some men from principle; it only has the effect on myself to let the writers and speakers see that, though not felt, it is not through want of perspicuity.

We will plant as nearly an entire crop as we have good seed, with the Cluster cotton seed; this is the original name, but known now by as many names as there are persons who desire to make money by selling seed. We will plant Silk (called McBride by some), 100-seed, Sugar Loaf, Dean, and small parcels of others. The Cluster, or Banana, has been much improved. The best now on sale is Boyd's Prolific. From this I have culled very carefully for three years, I think, and, by way of keeping solely for home use, I call them Home-seed; many, who have seen this selection, deem it better than the original accidental variety, for I learn from Mr. Boyd that it was an accidental stalk. Silk is perhaps better for all descriptions of land; many of my friends prefer it to Banana, objecting to the latter on poor, and on rich fresh land; on the first, the forms dry up; on the latter, breaks down—this latter can be remedied by topping, say one to two feet off. Sugar Loaf is best upon new ground, rich, sweet gum land. I have made over 41,000 lbs. the first year land was cleared, from twenty-four acres of land. 100-seed still retains its position on rich, fresh land.

6

Of the Jethro, several have inquired of me if I knew it. I reply: In the winter of '46–7, Col. H. W. Vick sent me eight small parcels of cotton in the seed, and asked my examination and experiment. They were endorsed thus: 100-seed, Lintonia, Diamond, Original Stock, Seed taken at random from a pile, Belle Creole, not a distinct variety, but inclining to Silk, eight locks of the Small Diamond, very valuable, Sub Ingri. These were planted April 23, 1847, hoed and picked by myself; no one permitted to touch, except ploughing. From the seventh variety (inclining to Silk), I selected what I deemed best in the lot. I sent a few seed to J. H. Hammond, ex-Governor of South Carolina, and to J. V. Jones, of Georgia. The latter brought it into notice, and I named it, in compliment to him, Jethro.

The history is comprised in a line. Col. Vick sent me a few seed, not half-a-pint; I planted and worked the crop. J. V. Jones made it tell. To the latter is due the credit, and so let it rest.

The Dean cotton was sent me about five years ago, I think, as Santa Maria, by a warm and devoted friend to agricultural improvement, C. B. Stewart, of Texas, from whom I have received many kind, similar favors; the production was so meagre that I discontinued the culture. After it had attracted attention, by fifteen and sixteen cent. price, he again sent me some, and Milton Cabeen, a personal acquaintance and friend, procured me a few seed from Mr. Dean himself. The yield is not one-half of my Banana, but the staple is excellent. Having been so unfortunate as to make all my crop ordinary to low middling, and getting some seven to eight cents, I conclude to make a better article, and now plant all my Dean seed. Silk and Banana yields about the same pure gin stand—say thirty-one per cent.—ginning out 500 to 1000 lbs. These yield more lint than any other variety I have tried.

One word as to selecting seed. A contributor of yours from Texas, is very correct as to the plan to be followed in making the best seed. It is what all planters should do, who desire to promote our cause. There is something else needed, and more than one in a hundred possess. Not alone the desire and care, but discrimination, judgment. We can tell that one article is not good, productive, &c., but to select the best is difficult. Frankly do I confess to want of that faculty, and have therefore preferred to rely upon the selection of others, and to labor to keep up that quality. It is sometimes good economy to buy a pair of pigs, even at $50, than to spend time and means to bring up to same perfection. This no one can deny. Why not the same of cotton, corn, oats, &c. It is the duty of every planter to strive to add to the knowledge and resources of our cause. We may fail, but the reward is sure—honest intention. Success to your efforts; may they be satisfactory to yourself, and a blessing to our land and nation. Yours, &c.,

Edwards, Miss., April 10, 1855. M. W. PHILIPS.

SECTION XIII.—COTTON SEED SPECULATIONS.

Mr. Editor:—I have noticed, in several of the last Nos. of the *Cultivator*, descriptions and recommendations of a variety of new sorts of cotton seed. I have also noticed this in the different papers of this State. There is, at this time, a greater variety of cotton seed in this State, than I have ever known; and I think that I may with safety say, that most of them were introduced for the purpose of speculating. In this, I will be condemned by many, but by only those who are engaged in it.

I can say, fortunately, that I am acquainted with some of the humbugs in the way of cotton seed. If you recollect, the Mastodon was introduced some four or five years since; and I recollect when there was not sufficient seed in this neighborhood to supply the demand at $5 per bushel. I am acquainted with the gentleman who first planted and sold the seed in this State; and it is generally believed that his profit was much greater from the sale of the Mastodon seed, than the proceeds of his entire crop for two years. At this time, there is not a seed of it growing, to my knowledge.

The Hogan, was the great cry last year; this year, I have heard but very little said of it. Some of my neighbors, who bought seeds at ten cents a-piece and planted them last year, will not plant them at all this year; which, I think, is sufficient proof that there is considerable of the humbug about them.

I believe that there is more fuss respecting the Banana, this year, especially in the county of Warren. I have never seen this article of seed; but, from what I can learn, they are identical with the Hogan and Pitt; and I think that the name has been changed in order to effect an increase in the sale of the seed.

The Brown, I think nothing more than an improvement on the Multiboll; and there is but little difference between the Multiboll and Sugar Loaf. These two kinds of seed I planted last year, and I find that they both yield well on fresh land, but do not do well on poor land. The only advantage that I can see, is in the picking. It does pick easier than the Mexican, but is much easier blown out of the bolls, and therefore more liable to waste.

But, Mr. Editor, I do not wish you to think that I am opposed to the improvement of cotton seed. I am as much in favor of it as any one, but my plan is to pick the Mexican seed. I think, or rather fear, that the introduction of the great variety of seed will ruin the Mexican. At present, it is almost impossible to get a genuine article of Mexican seed. By being careful, and picking our seed, we can improve them to a considerable degree. And I have noticed that those very men who are always collecting new corn and cotton seed, are the men that most generally fail in making full crops. I have no doubt but that a great many will readily conclude that I have been deceived in buying seed, and for this reason complain; but my reply is, that I never have bought or sold a cotton seed in my life, but I have some neighbors who are in the habit of trying all sorts.

I am, very respectfully,

Cayuga, Miss., March, 1850. HINDS.

SECTION XIV.—AGRICULTURAL HUMBUGS.

MESSRS. EDITORS:—When will humbuggery and extravagant representations of things and new discoveries cease? I will drop back about eighteen years, and bring up my subject as things may occur to my mind. Should I present names, I wish to be understood as doing it respectfully, not charging any one with willful false statements. Men's interest, in general, leads them, while giving an account of something new, to give a coloring to their statements that often leads their hearers into error or false notions This is what I wish to correct. Any one writing about seeds, or anything else, should first make fair and disinterested trials—proving the thing before saying everything in its favor. Always give the dark side, if it has any, as well as the bright. Get men's anticipations raised to a high pitch about a new kind of cotton, or any other seeds, and they, for the sake of gain, go in for the money, and soon find their disappointment, as in times of old—it sours, it will not keep—and cry out humbugged.

First, Baden corn, about eighteen years since, was run up to a high pitch by false statements, and men went in for seed at about $30 a bushel—a complete failure; the thing passed off in quick order. Next, in the year 1837 or '38, the Twin or Okra cotton seed came up; seed sold at various prices, from $5 a quart to $160 per bushel. It, by-the-by, proved to be a pretty good kind of cotton. I have been led to believe, that by a mixture of that kind with the Petit Gulf, &c., have, by proper selection and care, sprung up all those new kinds, or nearly so, that we have heard so much about for the last five years. Be it as it may, that seed has long been numbered with the things gone by. Next thing comes the making of sugar from common corn stalks. There was no speculation in this that I heard of; but what extravagant accounts did we hear? Some went so far as to make out that it would make more sugar per acre, with proper management, than the sugar-cane in its proper soil and climate. How has it turned out? Why, I suppose, a failure, as it has passed off—now never heard of. The Jerusalem Artichoke—not so much of a humbug, but overrated—has also gone by. Next comes Mr. Abby, of Mississippi, with his Mastodon cotton seed. I, fool like, went in for enough, one year, to plant twelve acres; and, as the devil said when he sheared the hog, "a great cry for a little wool." Mr. Abby could not keep the reputation of his seed up by all he could say in the papers. They had to pass off—though not before he realized several thousand dollars, as has been said. Next comes Remington, with his bridge and bed-slats—if not a humbug, certainly a failure, and has long since been laid as cold as a wedge. There is the Banana, Pomegranate, Sugar Loaf, Texan Burr, Silk, Brown, and Jethro cotton seeds—all very good, I have no doubt; but there has been more said in the papers, in the way of puffing, than was justifiable, it being to effect certain purposes, and that of a selfish nature. At last, the very kind for us Georgians has been accidently discovered by a Mr. Miller, of Mississippi. He styles it his Accidental Poor-Land Cotton. If this kind of cotton will do all Mr. Miller says it will, it would be a great misfortune for us cotton planters to have enough of the seed to plant full crops all over the cotton-growing country. He says it will make 300 lbs. per acre more than any other kind of cotton he is acquainted with. This kind of cotton, at this rate, would run a three million crop up to more than four millions, and this would reduce the price probably to four or five cents. Don't you see, Mr. Miller, that we had better let you keep and plant your seed? You say that you had rather plant your crop with them than take $1 a pint. My dear sir, $1 a pint for the seed is about $2,500 for the seed from one bale—about fifty times as much as the cotton sold for. Let us alone, friend, we are doing pretty well—we might do worse.

Next comes a new kind of corn—sprung up like Jonah's gourd, and for which I predict a similar fate. I can tell Mr. Ware that no kind of corn will do on common corn land, that bears from four to five stalks from one grain. But to the price—one and a-half gallons for $5, near $40 per bushel. This looks like doing brother planters favors. Next comes Mr. Young, with his superior kind of Yellow Corn; a very good kind, I admit, for strong land and good seasons. I have tried, I can't say how may kinds of corn, but have experimented enough to know that a medium-sized ear of either white or yellow corn is the best to be depended on, if planted in ordinary land and common seasons. I do not like this $2 a peck, when a good kind of corn can be had at thirty cents a bushel. I have a mind to say something for our much-respected old friend, the Doctor, of Mississippi. He has had a great deal to say in the papers about the different kinds of cotton seed, and, if I have not forgot, he, a few years since, spoke of the sale of cotton seed as being something in the order of a temporal saviour. He seems to be hauling off, and is very careless on the subject of the sale of cotton seed. You see what he says on the subject.

I might have said something of the Multicaulus fever, the Berkshire speculation, &c., &c., but have probably already said more than will be acceptable.

Yours, with great respect,

Atlanta, Ga., Feb., 1853. JOHN FARRAR.

SECTION XV.—SEA ISLAND COTTON PLANTING.

MR. EDITOR:—In the table on the opposite page you have the success of a Sea Island cotton planter for the last eighteen years, showing the amount raised per acre in each year, the price received per lb. for each crop, and the net proceeds per hand; also, for a part of the time, the appearance of the first blossoms, and the time of the first killing frost.

To the upland, and perhaps to the more successful Sea Island planter, I may seem to have been doing a very small business, still I think there are many who have not done any better; and as I know of no better way of measuring our success than by comparing notes, I shall be gratified in having the experience of any of my planting friends, for a longer or shorter time.

In the eighteen years my crops of cotton have averaged a fraction over three acres per hand, and a yield of 137 lbs. per acre, and net proceeds per hand, $83.

Liberty County, Ga., July, 1848. A SEABOARD PLANTER.

(TABLE ALLUDED TO ON PRECEDING PAGE.)

Year.	Yield of Cotton per acre.	Average price per lb.	Net proceeds per hand.	Time of first blossom.	First killing frost.
	lbs.	cts.	$		
1830	90	17	45		
1831	100	17¼	46		
1832	208	17¾	85		
1833	141	22	88		
1834	112	32	123		
1835	130	34	137	June 22	Nov. 28
1836	81	37½	87		
1837	85	26½	73	June 26	Nov. 23
1838	86	41	84	" 27	Oct. 30
1839	174	21½	95	" 19	Nov. 8
1840	154	27	110	" 10	Nov. 19
1841	153	15¾	61	" 18	Oct. 26
1842	223	13½	80	" 15	Nov. 11
1843	164	22½	91	" 22	Nov. 7
1844	146	14½	60	" 5	Oct. 29
1845	200	22	121	" 9	Nov. 10
*1846	68	24	41	" 16	Nov. 25
1847	156	14	70	" 7	

* Excessively wet, attended with caterpillars.

SECTION XVI.—SEA ISLAND COTTON—STATISTICS.

In a late Number of the Charleston *Courier*, we find a "*Report on Soils, Marsh Mud, and the Cotton Plant,*" prepared by Prof. Shepard, for the use of Mr. E. W. Seabrook, of Edisto Island. We publish it below, in the hope that it will prove interesting and useful to our readers on the seaboard of the Carolinas, Georgia and Florida; and prefix some valuable statistics upon the growth and price of the Sea Island cotton during twenty-two years prior to 1841. These statistics were compiled by the Charleston *Standard*, which we quote as follows:

We will add a few statistics, showing the value and importance of the Sea Island cotton crop. Extending our examination over a period of the twenty years preceding 1841, we find its production and price as follows:

1821	Quantity,	11,344,066	Av. price in	Liverpool,	21¼d.
1822	"	11,250,635	"	"	19d.
1823	"	12,136,688	"	"	17½d.
1824	"	9,525,722	"	"	19¼d.
1825	"	9,655,278	"	"	28½d.
1826	"	5,972,852	"	"	20d.
1827	"	15,140,798	"	"	14¾d.
1828	"	11,288,419	"	"	16d.
1829	"	12,833,307	"	"	15d.
1830	"	8,147,165	"	"	16d.
1831	"	8,311,762	"	"	13¼d.
1832	"	8,743,373	"	"	13¾d.
1833	"	11,142,987	"	"	16¼d.
1834	"	8,085,935	"	"	19¾d.
1835	"	7,752,736	"	"	24½d.
1836	"	8,054,419	"	"	25d.
1837	"	5,286,340	"	"	26d.
1838	"	7,286,340.			
1839	"	1,107,404.			
1840	"	8,770,669.			
1841	"	6,400,000—20,000 bales at 320 lbs. each.			

Since 1841, we have before us no reliable statistics, except with reference to the years 1850 and '51 and '52. With respect to the crop delivered up to the 1st of September in each of these years, it will appear that in 1850 it amounted, in the ports of Savannah and Charleston alone, to 26,634 bales, or 8,522,880 lbs.; in 1851, to 28,362 bales, or 9,075,840 lbs; in 1852, to 30,878 bales, or 9,878,900 lbs. And up to this date of the present year, we have 30,031, against 28,552 of the same time last year, giving us the reasonable assurance of a larger crop, by some 2,000 bales, than we have had for many years previous.

Nor is this the only improvement. The price has very greatly advanced, at least within the last year. The price at present ranges, for Santees and Maines, from fifty to fifty-five cents per lb.; for Floridas, from forty-two to forty-eight; and for Sea Islands, from fifty to seventy; and though this may be slightly above the ruling prices for the season, the average of all long staple cotton, for the entire season, would not vary far from forty-eight cents, leaving an immense profit to the planter over that afforded by any other staple. To pay as well as the short staple cotton, the long staple must sell for twice as much per lb. At present it sells for more than four times as much; and its cultivation must be, therefore, by so much the more profitable, and give by so much the greater inducement to its continuance and extension.

SECTION XVII.—SEA ISLAND COTTON PLANTING.

From the American Agriculturist.

Edisto Island, one of the largest of the South Carolina group, about thirty miles southwest of Charleston, containing 5,000 or 6,000 inhabitants, is the principal point where this valuable crop is cultivated. It is a sandy soil, but little above tide, which, flowing through many channels, gives very irregular shapes to the farms, but boatable water almost at every man's door. By this means the crop is conveyed to market, boats being substituted for wagons. There is considerable marsh, some of which has been reclaimed, and produces good cotton.

Salt-marsh mud is much used for manure, at the rate of about forty one-horse cart loads to the acre. Some compost it, others put it in the cattle pens. Some dry it before hauling, and then spread upon the land. Mr. John F. Townsend prefers to use it as soon as dug, spread upon the land wet, and ploughed in. He is the only man on the island who uses ploughs to any extent. All the land is cultivated with hoes, upon the two-field system; that is, one field in cotton, corn, and sweet potatoes; in the proportion of about seven-twelfths cotton, three-twelfths corn, and two-twelfths potatoes; in all, less than six acres to the hand. As the soil is generally very light, it is unproductive without manure. Therefore, as many cattle are kept as can be pastured upon the "field at rest," and the marsh and woodland. These are penned in movable yards, littered with fine straw and coarse marsh-grass or weeds, which is also used to lay along between the old rows, to which muck and manure is added, and all the grass sod which has grown during the year is hoed down into alleys, and the bed formed upon it, keeping the bottom as solid as possible.

If the plough were substituted for the hoe, twice as much manure could be made; or what, in my opinion, would be far more economical than digging muck or keeping so many cattle, merely to make manure, would be the use of guano. As this substance contains the same fertilizing properties of muck, in an hundred-fold degree, I would most earnestly recommend planters to try the experiment, by applying about 200 lbs. to the acre, ploughed in deep, or buried in the bottom of the cotton or corn beds. Make use of none but the best Peruvian, and purchase it from a reliable merchant, so as to be sure it is genuine.

It is true that cattle are easily kept here, living in winter in cotton and clover fields, eating the unmatured bolls of the former and stalks of the latter. In warm winters there is much grass, and in summer, I believe, it is rather abundant throughout all the south.

Cotton is planted from March 20th to April 10th, upon high beds, five feet apart one way, and from eight to twenty-four inches apart the other. Corn is planted about the first of April, upon the same kind of beds, from two to four feet apart. Sweet potatoes are planted the latter part of March; also upon the same kind of beds as the cotton and corn. As soon as the vines are sufficiently grown, say on the first of June, they commence planting the "slip crop." This is done by taking the vines from the seed beds, and laying along the top of other beds, and covering a part of the vines with dirt, when they immediately take root, and grow a better crop than from the seed. The bed is made rich and mellow, but the land below is kept as hard and firm as possible. The beds for cotton, corn, and potatoes, are all made in the same manner and distance apart, and are reversed every other crop; that is, changed into the alleys of the preceding one; but no rotation of crops is practised. The average yield of potatoes is about 150 bushels to the acre. Cotton, (long staple,) 135 lbs. Corn, fifteen bushels of the southern white-flint variety: no other will stand the depredations of the weevil.

The amount of labor to grow and prepare for market a hundred pounds of Sea Island cotton, is estimated at fifty days' work; that is, the small amount of labor which a negro does at "task work." The first process of preparing land for cotton, after manuring, is "listing;" that is, hoeing the grass off the old beds into the alleys. A "task" of this work is one-fourth or three-eighths of an acre a-day. Next, the old beds are hauled on top, at the same rate. The whole "task system" is equally light, and is one that I most unreservedly disapprove of, because it promotes idleness, and that is the parent of mischief.

The system of upland-cotton and sugar planters, of giving the hands plenty to eat, and steady employment, is a much better system. Meat is not generally fed to the laborers in this part of the State. The diet is almost exclusively vegetable, varying upon different plantations somewhat. The following are the weekly rations upon four places, which will give a general idea:

1st. One bushel potatoes a-week, from about October 1st to February 1st. Then one peck of corn, ground or unground, as preferred; or one peck of broken rice. Meat occasionally.

2d. One bushel potatoes, or ten qts. corn meal, or eight qts. of rice, and four qts. of peas, with occasional fresh meat, and twenty barrels of salt fish and two barrels of molasses during the year. Number of people 170.

3d. Half a bushel of potatoes, six qts. of meal, and about 2 lbs. of fresh meat, or ten qts. of meal, or ten qts. of rice. Carpenters, millers, drivers, and others who do not raise crops and hogs for themselves, have a much larger allowance.

4th. Half a bushel of potatoes, or ten qts. of meal, and at times, when the labor is hard, a quart of soup a day, and in light work twice a-week. This is made of 15 lbs. of meat to seventy-five qts. of soup, thickened with turnips, cabbage, peas, meal, or rice. Upon this place, as well as many others, the people can get as many oysters, crabs, and fish, as they like. They also keep a great many more hogs than their masters, but generally sell the pork instead of eating it. A half bushel of sweet potatoes, as measured out for allowance, by repeated weighing, averaged 43 lbs.

The process of preparing Sea Island cotton for market, after it is grown, is so remarkable, and so little known, that I will give the particulars.

In gathering it from the field, great care is taken to keep it clean, and free from trash and stained locks. Upon the drying scaffold it is sorted over, before packing away in the cotton house. When ginning, in fair weather, it is again spread upon the scaffold, and assorted. Some run it through a machine called a "trasher," that whips it up and takes out sand and loose dirt. It then goes to the gins, which are the same kind first invented; none of the many new inventions have been found efficient, and the Whitney gin totally unfit for Sea Island cotton. These simple machines are 3½ feet high, 2 feet long, and 1 wide, with an iron fly-wheel like that of a "box corn-sheller," upon each side, working a pair of wooden rollers, made of hard oak, about ten inches long and nearly an inch in diameter, held together by screws. In one instance, I saw a simple spring-bearer under the lower roller, and an iron one on top, to prevent the cotton from winding. These rollers wear out, and have to be replaced by new ones every day. I would recommend gutta-percha, as worthy of a trial, as a substitute for wood, as something tough and hard is required. The rollers are moved by the foot, like a small turning-lathe, the operator standing at one end of the gin, feeding the cotton very slowly through the rollers, leaving the smooth black seeds behind. A "task" is from 20 to 30 lbs. a-day, according to quality. Twenty or thirty of these little machines stand in one room; and, strange to say, none of those who have attempted to propel them by other power have succeeded. One very intelligent gentleman told me that he had spent $5,000 in trying experiments in machinery to gin this kind of cotton.

From the gins, the cotton is taken to the mote-table, where a woman looks it over very carefully and picks out every little mote or stained lock, as fast as two men gin. From the mote-table it goes through the hands of a general superintendent, or overlooker, and then to the packer. This operation is done by sewing the end of a bag over a hoop, and suspending it through a hole in the floor, and in this the packer stands with a wooden or iron pestle, packing one bale of about 350 lbs. a-day, as fast as it is ginned; as exposure to the air injures the quality, and it is not so saleable in square bales packed in presses, as it is in hand-packed bags.

The whole operation of preparing this valuable staple for market, requires the nicest work and careful watching of the operatives, as a little carelessness injures the value to the consumer. It is worth from 30 to 50 cents a pound—more than common wool.

The cultivation of these plantations is exceeding neat—too much so, probably, for the greatest profit, as has been proved, I think, by Mr. Townsend, in the use of ploughs instead of hoes. Mr. T. has also proved that sugar-cane will grow well, and has put up a small mill, and made some sugar. The cane matures fifteen joints, and granulates well.

CHAPTER IV.

DISEASES AND INSECTS DESTRUCTIVE OF THE COTTON PLANT.

SECTION I.—THE COTTON-WORM, ITS HISTORY, CHARACTER, VISITATIONS, ETC.

Correspondence of De Bow's Review.

THE following are some remarks on the nature of the cotton-fly of 1846, being a sequel to a dissertation on the usefulness of a knowledge of the natural history of insects, written last winter. I send you that portion only which treats of the cotton-fly, as falling more especially within the province of your periodical. This manuscript would not have sought a place upon your pages had not my attention been drawn to it by the ill-founded apprehensions of many planters concerning the *present* existence of the cotton-worm; an event utterly impossible, for if it makes its appearance at all this season, it most certainly will not do so until the cotton plant has attained its greatest maturity. I see also in your Review a communication claiming to show the means by which the army-worm may be effectually eradicated, in which is displayed the greatest ignorance as to the general laws which govern the insect world. The writer states that the chrysalis of the cotton-fly may be ploughed up, and thus destroyed, &c. Now these chrysalides never go in the ground at all, but are invariably attached to something above the surface. This is a fact that could not have escaped the attentive observer. I ask how a chrysalis invariably formed above ground, and incapable of locomotion, is to work its way beneath the soil? As to the insect in any condition secreting itself in the earth, beneath the bark of trees, under fallen timber, &c., it is altogether a mistake, if not an absurdity, and easier asserted than proved. In treating of the cotton-fly in the following pages, my aim has been to found my assertions upon general principles, and though the practised entomologist may find some inaccuracies in the detail, yet I insist upon the principles as universal and incontrovertible.

Let us now pass to the consideration of the cotton-fly, premising, however, before entering into an examination of this destructive little moth, that my remarks are intended less to enlighten others than to elicit information from some one who is better able to inform the public mind on this interesting subject. As for myself, I must confess that my limited observations do not justify me in coming to any positive conclusions, and have by no means satisfied my curiosity; but my information, such as it is, I give in the following pages, with the hope that, however imperfect it may prove in the main, yet that some mite of information may be gleaned from it. It is impossible to think for a moment that this species of moth has escaped the observation of entomologists, for the plant upon which it feeds to the absolute exclusion of all others, (being the great staple production of many countries,) must have brought it into notice at various times and at various places. From its univorous nature, (to coin a word,) it must have been coëval with and inseparable from the existence of the cotton plant. My principal motive for broaching this subject is on account of the frequent remarks made and fears entertained, that the army-worm would become an annual plague. But since I have investigated their nature I have come to the conclusion that these fears are groundless, and that the cotton-fly can never become naturalized in our climate.

The first irruption, as I am informed by an old planter, that this insect made on the cotton fields of Louisiana, was about the year 1820, when its progress was marked with the same utter destruction of the cotton crop as in the subsequent years of their appearance. It then disappeared until '40, a period of twenty years. There is something singular and unaccountable in the periods of this insect, something vastly different from the periodicities of others which we find with us, for they appear to be governed by some fixed laws; the most of them are annual, very few biennial. Now, the grasshopper, house-fly and mosquito may be looked for at the return of summer with as much confidence and certainty as we look for the revolutions of the seasons. The *cicada septendecem* never fails to make his appearance once in seventeen years. But who can tell whether the cotton-fly will appear next year or fifty years hence? No scourge, whether under the form of a devouring insect or that of a malignant disease, ever became annual in one particular place. Look at the locust of Egypt; suppose that voracious insect to become annual, the prolific valley of the Nile, once the granary of Asia and Europe, would become a howling desert. Look at the plague that devastates sometimes Smyrna and Constantinople; did the cause of that distemper act with the like intensity at each return of the season, those flourishing cities would long since have been numbered with Thebes and Memphis. Let the cholera or yellow fever prevail in New Orleans every year, as it has at times, and that great emporium of the Southwest would become a puny village. Is there not an invisible hand that sways the destinies of the world? a hand that stays the devastations of plague, pestilence and famine? The cotton-fly belongs to that numerous class of insects known to naturalists under the term of *phalena* or moth tribe. The following are its specific characters, without the technicalities made use of by the naturalist, so far as they could well be avoided. Antennæ, or little horns projecting from the head, setaceous or terminating in a point like a bristle, of a drab color, five lines in length, being about half the length of the body. Wings incumbent, deflexa; under surface of thorax or breast of a dull silvery white, insensibly terminating on the abdomen, and wings in a color tending to a russet; the upper surface of the wings and back varying somewhat in different individuals, but generally of a changeable golden color with ferruginous zigzag lines traversing the surface transversely; posterior margin bordered with a narrow strip of pale pink color, with small denticulations. On the upper surface of the wings there are two black spots, one on each, about the middle of the base; legs white, the four posteriors very long when compared with the front ones, which are short and slender; the tail simple. The length of this insect is about nine lines from head to tail. Expansion of the wings, at the tips, about the same measurement. To conclude, I will add that the shape of this moth is very much like that of an isosceles triangle, with the line forming the base inflected inwardly about two lines. This peculiar figure is produced by the exterior angle of the upper wings projecting beyond that of the interior angle.

During the present year, the time that my observations commenced for the first time, the cotton-fly again made its appearance in the latter part of August, at first making but little progress, but about the middle of September their numbers increased so prodigiously, that in many instances they would eat over a field of several hundred acres in four or eight days. The number of eggs deposited by the female is uncertain; they are smaller than a mustard seed, and always deposited on the under surface of the leaf during the night; in a few days their eggs hatch. The worm, at first a minute living point, falls immediately to work to devour the leaf; its growth is rapid, for its labors cease not night nor day until it arrives at maturity; it then winds itself up in a leaf by means of a web resembling a *cobweb*, casts its skin and changes into a chrysalis, in which state it remains ten days, then it bursts the thin walls of the chrysalis, and comes forth a perfect insect. In turn, it begins the work of reproduction, deposits its eggs, and in ten more days it dies.

Thus in every ten days there is an additional generation, and they go on increasing *ad infinitum*. As soon as the leaves were consumed in a field this great *army* took up its march: some in search of comfortable quarters, where they might repose from their labors; others on a foraging expedition to replenish the means of their subsistence. The first took shelter in the first leaf they met with, but generally they proceeded as far as the fence, a barrier beyond which they never travelled, where they found a plentiful supply of leaves, in which they enveloped themselves. The second division extended their march much farther, sometimes travelling half a mile from the point whence they started, perishing by cart-loads for the want of food and the many casualties to which their journey subjected them, such as carriage wheels, heat of the sun, and the rapacity of birds.

Here then it would appear was an end of the cotton-worm for a season at least; for those which yet remain in chrysalis in the fence-corners, will change to the fly in ten days. But where are now the cotton leaves upon which the pregnant female is to deposit her eggs? There is not one left. If they are placed on any other leaf the eggs may hatch, but the worm must perish, as we have just seen them perishing by myriads while wending their way through a various and luxuriant herbage in search of that food intended for them by nature. In ten days from the time that the worm becomes a chrysalis on the borders of the cotton fields, a host of flies are seen issuing therefrom: they go forth in search of food for their forthcoming progeny; now it is to be found their days are numbered, in ten more if they meet with no cotton leaves, they themselves must die, and thus put an end to the whole race. But their search is continued, and now, when the weary insect is ready to finish its term of days, a tender but sparse foliage crowns the leafless twigs of the cotton plant. On them the eggs are deposited: they hatch, the worm eats, returns again to its chrysalis. The cotton stalk still puts forth new leaves, they grow and expand until the fields again look green; ten days, ay, forty elapse, yet there is not a worm to be found. One would have thought that this second crop of leaves would scarcely have been sufficient for a single repast for them, yet the food that they so lately devoured with such voraciousness is now left untouched. What is the matter? Why don't they eat, their food is spread before them? Read on, the answer will be found in the sequel. Let us examine the cause. In nearly every fourth leaf we find a chrysalis writhing and contorting itself at the touch. Ah! here is the explanation of the difficulty; this is no ten days' chrysalis, but that in which it is to hibernate, possibly for one winter, perchance for twenty. Let us take a pocketful of these home, and place them beneath tumblers, and wait patiently to see what they will produce. If I had found a treasure my delight could not have been greater than that I experienced at the idea of unravelling this mystery. But man is prone to disappointment, as we shall soon see. About the 15th of November, the insect appeared, but *mirabile dictu!* as different from the cotton-fly as it is possible to suppose one insect could differ from another. It belonged altogether to a different family, a description of which I give, as follows:

Antennæ filiform; black, six lines in length. Palpi four; two external and two intermediate, the external white, twice the length of the other two, in shape angular, the angle projecting externally. The two middle are straight, scarcely perceptible over a strong light; they are of a dark color. Wings four; hymenopterous; incumbent, extending to and exactly even with the end of the tail; shape of the wings, which are small and extremely thin and delicate, like that of a fan. Front legs half the length of the posterior, of a uniform orange color; the intermediate legs very little longer than the anterior; the thighs of a deep orange color, the rest of the leg annulated with orange and white. The posterior legs long in comparison to the others; thighs of a deep orange color, the rest of the leg annulated with black and white, the rings being larger than those of the intermediate. The trunk is a uniform shining black, as would be the upper surface of the abdomen also, were it not for the very narrow white bands which connect the black scales together, giving to the abdomen an annulated appearance; these white lines do not encircle the abdomen, but terminate uniformly on the sides. On the under surface of the abdomen these white rings again commence, which are much larger than those on the upper surface, causing the abdomen to look almost white. The tail terminates in a bifurcated sheath, enclosing a long blunt sting, projecting considerably beyond the tail, and forming a very prominent feature in the general figure of the insect. This is a small, slender insect, much longer than the honey bee, but not so thick.

Now it is evident from its specific character, as well as from its parasitic nature, this insect belongs to that numerous class called *ichneumons,* of which there are upwards of five hundred species. As I am not at present in possession of any practical work on Entomology, I cannot determine the species of this ichneumon, but to show that it differs in some respects, from the family to which it belongs, I will quote a paragraph from a work before me, in which are set forth some peculiarities belonging to that class of insects as a genus :

" The whole of this singular genus have been denominated parasitical, on account of the very extraordinary manner in which they provide for the future support of their young. The fly feeds on the honey of flowers, and when about to lay her eggs, perforates the body of some other insect or its larvæ with its sting or instrument at the end of the abdomen, and then deposits them. The eggs in a few days hatch, and the young larvæ, which resemble minute white maggots, nourish themselves with the juices of the foster parent, which, however, continues to move about and feed until near the time of its changing into a chrysalis, when the larvæ of the ichneumon creep out by perforating the skin in various places, and each spinning itself up in a small oval silken case, changes into a chrysalis, and after a certain period they emerge in the state of complete ichneumons."

It will be seen that there is a peculiarity attached to this ichneumon not included in the above description ; that of appropriating the chrysalis, as well as the larvæ of other insects, to the use of their young. All ichneumons that I ever read of, spin their own chrysalis, but this is the prince of parasites, for not content with eating the substance of his neighbor, he seizes also on his house. So far as I have read concerning this curious family of insects, this is a nondescript. As an example of these insects called ichneumons, I may mention the *ichneumon seductor,* or dirt-dauber; well known to everybody as that wasp-like insect, which builds its clay houses on the walls, and particularly in the recesses of windows, to the great annoyance of the tidy housewife.

Thus is answered the question, why the cotton-fly did not again eat up the scant foliage which subsequently appeared on the stalks. This little usurper goes forth in search of "whom he may devour," and as soon as he finds a house built and well provisioned, he seizes upon it for his posterity, which he does in the following manner: When he finds a cotton-worm, he pierces it with the instrument with which its tail is armed, and deposits an egg; the cotton-worm soon spins itself up into its case, there to await the period of its perfection, which never arrives, for soon the egg of the ichneumon hatches, and falls to devouring his helpless companion. This work of extermination continues until there is not a vestige of the cotton-fly left. I venture to say, while I am now writing (1st of December), there is not an egg, chrysalis, or fly, in the confines of the United States. My speculations on the nature and habits of the fly have led me to adopt the following hypothesis: That it is a native of tropical climates, and never can pass a single winter beyond them, consequently, never can become naturalized in the United States, or any where else where the cotton plant is not perennial, for nature has made no provision by which they can survive more than ten or twelve days, therefore they must perish wherever the cotton plant perishes during a period of six months. That wherever they have prevailed in our cotton-growing regions, it is when they have become very numerous, and consumed all the cotton in their native climes, and then go in search of their food in more northern climates. It is not to be presumed that this happens often, but the same remark will hold in regard to the cotton-fly as it will to many other insects, that owing to some unknown cause, they become exceedingly numerous, but at long and irregular intervals. The locust has already been noticed as an example, and many more might be cited. I, however, will mention another to which I was an eye-witness. About eighteen years ago, the *green* or *blow-fly* became so numerous that thousands of animals perished by them, also some human beings. The least spot of blood, the moisture of the mouth, eyes or nose, was sufficient to cause a deposit of eggs. Sick persons, particularly those who had not proper attention, suffered. Several negro children, who came under my notice, fell a sacrifice to them; and it was with difficulty that many others were saved. In these instances, the fly deposited the eggs within the nostrils, where they soon caused death by producing inflammation of the brain. This fly is annual, and scarcely ever deposits its eggs on an animal, except it be the victim of a running sore; but at the period alluded to above, it appeared that there was scarcely animal flesh enough to feed the maggots of this numerous host. It is but once within my recollection that I have witnessed this phenomenon; and neither before nor since have I heard of such ravages of the green-fly. Why they should have existed in such incredible numbers at the time referred to, is a question not to be easily answered.

There are three circumstances upon which I found my arguments in support of my hypothesis of the cotton-fly: First, Nature has made no provision by which it could survive the winter season. Second, The irregularity of their appearance. Third, Their progress from south to north, and from west to east.

It may be remarked, on proposition first, that all insects included within the genus *phalena,* hibernate in the state of a chrysalis, therefore it is utterly impossible for the cotton-fly to hibernate in that manner, as they remain but ten days in chrysalis. The fly does not hibernate, for the period of their existence is but ten or twelve days. It cannot be in the state of the egg, for it is a law equally inflexible with regard to this tribe, that the egg must be deposited on the leaf on which the larvæ are to feed; and the reason is very plain, for these larvæ, when first hatched, are minute living points, of an exceedingly helpless nature, almost devoid of locomotion, or possessing it in too small a degree to enable it to go in search of its food. But let us suppose that the egg does survive the winter; how does it happen that when the worm first makes it appearance, it is found on the very summits of the cotton, instead of the lower branches? parts that it would reach the soonest, if it proceeded from the ground upwards.

The *phalena mosi*, or silk-worm, is an insect of the same genus as the cotton-fly, and whose habitudes are very much the same as the latter, tropical in its nature, confining itself to a particular vegetable, the different species of mulberry, and being short-lived in the chrysalis, remaining in this state but fifteen days. At the approach of winter, when the mulberry trees cast their leaves, and remain leafless for many months, these insect, in our climate, would all perish, were they left to themselves. But art, in this respect, has triumphed over nature; for the silk grower at a certain season gathers a parcel of eggs, and places them in a cold dark place until the mulberry tree shall again afford them food in the spring, and in this manner they are perpetuated, and this is the only possible way that they could be preserved here. They are like some tender exotic, which flourishes as long as the warmth of the hot-house affords them a congenial atmosphere, but perishes if left to buffet the rigors of winter.

Proposition second. Here I contend that when an insect is a native of, or naturalized, in any country, they are always governed by some invariable laws which determine their appearance. The grasshopper is annual, coming every spring or summer; the locust of our climate septem-decennial, appearing once in seventeen years; but the cotton-fly has no regular periods of return, showing that when it reaches our climate, it is by some casualty.

In proposition third, I maintain that if the cotton-fly sojourns here during the winter or winters, when it did appear at all, it would do so simultaneously through the whole cotton district, instead of which we see it progressing regularly from south to north, and from west to east.

Such are the speculations that I have entertained concerning the cotton-worm, from which I conclude that it originates in South America, and reaches us through Mexico, and never can become a denizen of our soil.

Bayou Sara, June 1, 1847.

SECTION II.—THE RUST AMONG THE COTTON.

From the American Cotton Planter.

DURING my geological tour through some of the eastern counties of our State, I have frequently (more especially in the prairies,) been asked, "If I could suggest a remedy for the rust on the cotton plant." I have invariably stated my views to all those who inquired ; but as that question is one of general interest, it will not be amiss to repeat them here, on the pages of a popular journal, devoted to southern planting and farming.

The boll-worm and the rust are decidedly the arch-enemies of the cotton plant; and I am very much afraid that the first will, in the course of comparatively a few years, increase to such an extent as to destroy two-thirds of the whole cotton crop, render the cultivation of that plant unprofitable, and ruin in that way our southern El-dorado. A remedy for this scourge of the cotton planter, which now destroys fully one-third of our cotton crop, although it made its first appearance not more than ten or twelve years ago, can, with our present ignorance about the nature of that insect, possibly not now be found. In the interest of the southern planter, I have tried to draw the attention not only of the agriculturists of the southern States, but also that of Mississippi, to that all-important subject; and prominent southern editors, of this *Cotton Planter*, and *DeBow's Review*, have kindly assisted me in promulgating my suggestions; but so little have I succeeded in arousing the attention of those societies, that my disinterested communications have not even elicited a satisfactory consideration and answer.

If a remedy against the increase and ravages of the boll-worm cannot now be devised, on account of our ignorance of the nature of that insect, the case is different with the rust of the cotton plant. The nature of this rust is easily found out by the aid of a sufficiently powerful microscope, and known to be nothing else but a parasitical fungus, growing upon the stock and branches of the cotton plant. This fungus is produced by a diseased state of the plant, caused by a stagnation in its growth, and a consequent relaxation in the circulation of the fluid or sap of the plant. Such a stagnation in the growth of the cotton plant can be produced by an unfavorable season, it is true, and rust will appear in such cases everywhere, even in the freshest and best kinds of soil. Such cases are beyond the control of the best agriculturist, and belong to those chances which he has to bear; but such cases are extremely rare—of one hundred cases of rust among the cotton, perhaps scarcely one is owing to an unfavorable season, and ninety-nine to a defective cultivation; and these cases are consequently under the control of the agriculturist.

The rust appears only very seldom on fresh land; but most generally on such as has been for some time under cultivation, and is exhausted by abuse, or an unnatural or defective management. Agriculture is, in our southern States, not yet carried on as an *art and a science,* which it is indeed, but, unfortunately, only as a mechanical business, which we continue to execute in that rude manner as it has been handed to us by our ancestors, and modify it only according to our convenience. We ask every thing from Nature, and are unwilling to do more than is absolutely necessary. The unavoidable consequence is, that in a very few years we exhaust the best of our lands; they then refuse to yield adequate crops, and produce diseases of the vegetables which blast our hopes.

A plant does not only draw its food from the atmosphere by means of its foliage, absorbing the oxygen, nitrogen, carbon, and ammonia, which the atmosphere contains, but it requires also a certain amount of the inorganic constituents of the soil in which it roots; it lives, therefore, mostly at the expense of that soil. Every plant requires always the same nourishment; the consequence is, that if the same plant be cultivated or live a long time on the same soil, it must, in the course of time, be entirely deprived of those substances which that plant requires for its growth. May I illustrate this by an example?

According to an analysis of the ashes of the cotton plant, (made in one of the northern colleges, and which I give for what it is worth,) it contains, in 100 parts:

1. Potash, - - - -	29.58
2. Lime, - - - - -	24.34
3. Magnesia, - - - -	3.73
4. Chloride, - - - -	0.65
5. Phosphoric Acid, - - -	34.92
6. Sulphuric Acid, - - -	3.54
7. Silica, - - - - -	3.24

Potash, phosphoric acid, and lime, are therefore the principal ingredients of the cotton plant; and, in order to live and succeed, it must consume them in sufficient quantities. The atmosphere contains none of those contents, consequently the whole amount of them must come from the soil in which the plant roots. If, now, the cotton plant be cultivated for a number of years in succession on a body of land, that land must be entirely deprived of those ingredients, at least of that part which is in solution. As soon as the quantity begins to become insufficient, even only of one of those ingredients, the cotton plant will no longer grow vigorously; a stagnation of its growth must take place until the insufficiency has been supplied; such a stagnation produces diseases of all kinds, and among them the rust.

A sovereign remedy against the rust is, therefore, the introduction of a system of agriculture in conformity with Nature, and with the science which has been abstracted from a long practice, and the investigation of the nature of plants and soils by chemical analysis.

If we observe Nature closely, we will find that if certain vegetables, which have grown for a long time upon the same soil, are removed from that soil, the same kind will not be reproduced spontaneously, but quite different genera and species will appear; the simple reason is, because the soil has been exhausted of such ingredients as that kind of vegetables require for its growth; but there are still other ingredients in it which are suitable for other plants, of a different genus and species, and such will appear spontaneously upon the soil, and grow luxuriantly. Nature points, therefore, to a rotation of crops in agriculture; and if we obey Nature and observe such a rotation—if we supply, from time to time, those ingredients which are most necessary for the growth of our crops, our lands will never be exhausted; on the contrary, they will improve, and the vegetables which we cultivate will grow luxuriantly, without a stagnation in their growth,—they will remain free from all those diseases caused by such a stagnation, consequently also from *the rust*, and yield an abundant crop.

Not having been accustomed to systematic manuring of our land, we think it very difficult, laborious, and even expensive. It is indeed not so; it is much less troublesome and expensive than clearing and taking in new land when the old is exhausted and unfit for further cultivation; but we are accustomed to the latter, and not accustomed to the former, therefore we are prejudiced against it, and imagine it to be much more troublesome than it really is. In some of our States it is difficult to procure manure, and some trouble may arise from that circumstance; but, indeed, if we introduce a manner of agriculture suitable to our soil and climate, much manure is not required to keep our land in constant fertility. The first requisite for such an agriculture, is to prepare our soil well, to plough and harrow it in such a manner that the soil becomes perfectly mellow, to allow the vegetables which we sow or plant in it to take root, and the atmosphere to operate upon it and have a dissolving influence upon its contents.

If we sow our wheat upon entirely unprepared land, from which the corn or cotton has just been harvested, without ploughing and harrowing it before, and bury the seed with brush-wood, as the uncivilized Indian would do, unprovided with the implements of enlightened agriculture, it is of course impossible to reap an adequate and remunerating crop; it may perhaps be adequate, not to the forces of our land, which have not been developed, but only to the rude and imperfect manner of our agriculture. Land cultivated in such a manner has neither been exposed to the dissolving influence of the rays of the sun and the atmosphere, nor has it been made mellow enough, and a few grains of the seed will only fall in such a situation where they can germinate and root easily; the necessary consequence will be, that the wheat stands too thin, but even for the thin stand the soil is not sufficiently worked, and its forces not developed; diseases of all kind of the plant will appear, and the crop be a very poor one. Such an agriculture is nothing but a rude attempt to save labor, but it is only done at the expense of the crop, and if labor be saved, the value of the crop is certainly most unproportionably diminished; hence it happens then that even from fine land, which ought to yield from twenty-five to thirty bushels of wheat, an average crop of from six to twelve bushels can only be made. Let me ask if a little more labor would not be well paid, if, with that additional labor, consisting of ploughing and harrowing the land before we sow the wheat, and then again harrowing the seed under, we can more than double the crop? The reward will certainly appear to be an ample one, if we consider that, with such a manner of agriculture, half the land is sufficient to produce the same quantity of wheat as the double quantity with a rude and imperfect cultivation.

Our cotton land is generally better prepared than our land for small grain, but, by its cultivation, we commit especially the grave error, to continue for a number of years to plant the cotton in the same land, instead of introducing a rotation of our crops. With such a rotation, a little manure is all-sufficient to keep the land always in a fine state of fertility, and to improve instead of exhausting it. Let us see how such a rotation can be most profitably managed:

Our crop here in the southern States consists, principally, in cotton, corn, wheat, oats, peas, or small grain in general, and potatoes. Half of our land, at least, is generally planted in cotton, and the other half in corn, small grains, and potatoes. Commencing now with a fresh body of land of one hundred acres, we may plant the first year fifty acres in cotton, twenty-five in corn, and twenty-five in small grains and potatoes. The second year we plant the cotton on those fifty acres from which corn, small grains and potatoes have been harvested, and, *vice versa*, the corn, small grains and potatoes on the cotton land. The third year we plant the cotton again on the same fifty acres on which it grew the first year—the corn on the land on which the small grains and potatoes grew the first year, and the small grains and potatoes on the land which produced corn the first year. Every part of the land has then borne in succession all the different vegetables of our crop, and half of it has borne our principal article, cotton, twice. The fourth year we let twenty-five acres of this last-mentioned land rest, and fallow it, and divide the remaining seventy-five acres into three parts, viz.: thirty-seven and one-half acres for cotton, eighteen and three-fourths for corn, and eighteen and three-fourths for small grains and potatoes, selecting the cotton land from those fifty acres which have only once borne a crop of cotton. If we observe such a change, and fallow one-quarter of our land every year, so that after four years every portion of the one hundred acres has had a year of rest, we can cultivate our fresh land, according to its quality, from eight to twelve, perhaps from sixteen to twenty years without manure, and we will, with such a rotation, scarcely perceive any decrease of its fertility. After the lapse of those years, we not only keep up its fertility, but even increase it, if we manure those twenty-five acres which have been fallowed. After the lapse of four years, all our land has in that way been manured. Such a system of agriculture is easily introduced; and followed up with care and without much labor and expense, it will not only not exhaust the land, but increase its fertility; it will render agriculture more easy, the land becoming more and more mellow and disintegrated and deprived of those stumps of trees and roots which obstruct the cultivation of the soil so much; and lastly, it renders the taking in of new land perfectly superfluous, unless we increase our workers. The continual fertility of the land will prevent a stagnation in the growth of our crops; there will be little or no disease among them, and especially the rust of the cotton plant will seldom appear, and only then when the unfavorable seasons produce it.

Such a rotation of our crops has another most salutary and remunerating influence upon our cotton field—*it will most certainly diminish the ravages of the boll-worm, and the enemies of the cotton plant in general.* The boll-worm is a caterpillar, the larvæ of a lepidopterous insect or butterfly of the night-swarming family, called Noctua, which, as all the insects of that tribe, undergo, after having been hatched, three distinct metamorphoses, or changes. The insect originates in the form of a small egg, not near as large as the head of the smallest pin; the hatching of this egg, after a few days, produces the worm or caterpillar; this, when full grown, changes into a chrysalis or cocoon, and this, after ten or twelve days, is transformed into the perfect insect, butterfly or noctua. The individual natural history of the boll-worm is as yet very little known, but having the generalities of its nature in common with other insects of the same tribe, which are better known to entomologists, it must be, during the winter and the whole time when there is no food for it, either in the state of an egg, which is indeed most probable, or in the state of a chrysalis or cocoon; it can possibly not hibernate as a perfect insect or butterfly, not finding any food until late in summer. The eggs or cocoons that hibernate must be hidden in the neighborhood where the perfect insect lived, consequently in the cotton fields or near them. If such fields are not planted again in cotton next spring, the largest number of the brood must necessarily perish, the little caterpillars not finding any aliment suitable for them, and not being able to creep to other cotton fields. A generally introduced rotation of crops must accordingly greatly diminish this dreadful enemy of our principal southern staple. We perceive, therefore, how beneficial such a rotation of crops must be; it not only prevents the most pernicious diseases of the cotton plant, but also the ravages of its most dangerous enemy, the boll-worm, and will certainly save us the one-third of the whole crop.

In our prairie soils, and wet and heavy soils in general, there is another cause which produces the rust among the cotton; this is the superabundance of moisture and the stagnation of the rain-water in the field. It is this which renders the prairie soil especially subject to the rust of the cotton plant. Too much moisture and stagnant water, heated by the rays of the sun, produce immediately a stagnation in the growth of that vegetable; it does not allow it to imbibe enough of that solid matter necessary for its growth, especially as this plant is much more adapted to dry and light, than to wet and heavy soil; if we, therefore, will plant cotton in heavy and wet, especially in prairie soil, it is *absolutely necessary* that this soil should be as much as possible protected against superabundance of moisture and stagnation of rain-water. This can only be done by a vigorous system of draining; by ditching where it is necessary, and leading the water, by means of deep furrows, into the ditches. In fact, in no soil a system of ditching is more necessary than in the prairie soil. If it is neglected even only in one place, and the rust makes its appearance, if only in that one place, it will soon spread over the largest portion of the field, it being an infective disease; the minute seeds of the microscopic mushroom ripen quick, and are carried by the slightest breeze all over the field.

University of Mississippi, Dec. 23, 1854. L. HARPER.

SECTION III.—INSECT PHYSIOLOGY—THE BOLL-WORM.

MR. EDITOR :—I have concluded to write you an article or two on the insects which are injurious to the agriculturist of the South. I will begin with what is vulgarly called the *boll-worm*, a caterpillar, which, for the regularity of its visits and length of time it remains, we may consider as fixed upon us. This is decidedly the most destructive insect that feeds upon the cotton plant in this climate. Insects of some sort prey upon almost every species of the vegetable kingdom, and we must learn the habits and natural history of insects, if we wish to discover the most effectual remedies to prevent their depredations. This insect is an anomaly in the natural history of insects, for it is much more destructive to the plant, cotton (*gossypium*), for which it was never made, than to the one to which it naturally belongs, corn (*zea mays*). If I am right in my supposition, this insect is the caterpillar we find in the end of ears of corn, eating the silk, and some little of the corn. This insect is at the north as well as at the south—in fact, it is wherever the corn grows, and will never depredate upon the cotton plant, unless through necessity. The moth of this caterpillar belongs to the order *lepidoptera*. The character of this order is (according to the system of Dr. Leach) wings four, covered with scales, tongue spiral, filiform. Linnæus divided this order into three generations, *papileo* (butterfly), *sphinx* (hawk-moth) and *phalæna* (moth), which were characterized by the form of their antennæ. Genus Phalæna, antennæ moniliform, shorter than thorax, palpi very small and very hairy. Wings elliptic, equal, long. To this genus belongs the group *agrotididæ*, the larvæ of which lies concealed in the ground, and feed by night (as the cut-worm); and the group, *mamestradæ*, the larvæ of which lives exposed, and transform in the ground, as the cabbage caterpillar. The insect I will call Phalæna Zea (corn-moth) until it is more correctly classed, belongs, perhaps, to the latter group.

The P. Zea, or corn-moth, is of a pale yellow or a shining ash color—length of body and wings one and one-eighth of an inch, the wings expand two inches horizontal, the upper wings covering the lower; below the centre and near the border of the upper wings, are two dark spots; there are some two or three indistinct ones on each upper wing, end of the wing whitish, a wavy dark band near the border. Thorax slightly convex, downy; abdomen color of wings, downy; proboscis folded spirally underneath, double, half-inch long; eyes large, clear, yellowish green. Legs six, antennæ fusiform, palpi very hairy, flies only late in the evening and at night, lies concealed in the day in jams of the fence, around stumps, and in the grass and weeds, flies rapid and low.

The Maize Phalæna pairs with its mate as soon as found (some insects of this order have a remarkable instinct that way); the moth lays about 750 eggs, on the fourth day, about the size of cabbage seed, of light cream color, and dies in three or four days afterwards. The moth sucks the nectar from the bloom, or rather between the calyx and petals. In confinement, they will suck water sweetened with sugar. The eggs of the first brood are laid on the silks of corn; if no silks, on the top of the corn; you may very often find them in the northern corn we plant for early roasting ears. The ova or egg will hatch in two or three days. The larvæ feeds upon the silk and the grains of the corn, remains in the ear for fourteen days, comes out and goes into the earth about three inches, and is transformed into a chrysalis of bright, shining mahogany color, conical in shape, seven-eighths to one inch in length; it remains in the ground from fourteen to sixteen days, when its second transformation takes place, and it comes out the moth I have above described.

The second brood comes out from the 15th of July to 10th of August; it now finds but little corn to go to (at least in its section of country), and necessity compels it to deposit their eggs on the cotton plant. Their eggs are laid on the top bud, and the bud of the end of the limbs; sometimes, when very numerous and late in the season, on the leaves promiscuously. If at the time of this deposit the weather is dry, and the sun very hot, the ova or egg becomes abortive. Hence the phrase, " no worms of a dry year."

However, during the hottest and dryest season, enough will escape to do some damage. Thousands of the eggs and young larvæ are destroyed by ants, and the ichneumoniadæ. The larva spins around it a thin web, when first hatched, for protection from ants and other enemies, and will swing itself by a thread, if it fall from your hand when first hatched, say five or six days old—sheds its skin until eight or ten days old —it descends from the tops of the cotton and the ends of the limbs in two or three days after being hatched out, begins its depredations on the forms by eating through the calyx in the petal (so small is the place that you can hardly discern it), which makes the bracts or floral leaf turn yellow, and the form falls off; the larva does not wait for this, but is off to another and to another, until it destroys four or five, when it comes to a boll into which it goes and lies concealed, if enough to feed on, until the usual time of its transformation. The caterpillar is sometimes killed by hot sun, while eating into a boll.

If we have a short season, we will perhaps have but two broods. This is the case in Tennessee, and sometimes in North Alabama. The year 1848, I made a good crop of cotton, but it was made after the disappearance of the caterpillar. I cannot account for their disappearance, for the season was favorable. They may have been destroyed by some of the ichneumoniadæ family, perhaps the white oblong dots we saw on them. I never saw them on first brood or their eggs. But this is all hypothesis.

Another reason why they do not damage the Tennessee planter so much is, that he plants and grows corn all the season, and the moth lays her eggs on corn in preference to cotton. We will see the difference between two broods and three. Say you have 200 moths to come out, one-half are males: we take 100 females at 700 eggs each, say 70,000 caterpillars the first generation; 24,500,000 the second; now sum them up to the third, deducting half for males, and we have the enormous sum of (if I have not miscalculated) 8,575,000,000. This insect hybernates in the chrysalis state in the ground.

The larva or caterpillar, when full-grown, will measure from one and one-half to one and three-quarter inches in length, it looks to a superficial observer brown, pale yellow and light green, though it has eight longitudinal streaks of white, brown and green, with one or two dots on each segment of the body along the lowest streak; it is smooth, shining, naked, with a few hairs on each segment of the body. They are of a cylindrical form, tapering a little at each end, rather thick in proportion to their length, legs six before, eight central, and two anal. Head brown, smaller than body, oval. I know of no effectual means of preventing the ravages of this insect, but that the remedy is worse than the disease. Now, if we were to plant no corn (*zea mays*), we might get entirely clear, perhaps, of this insect; but more anon.

Jackson, Miss., July, 1850. JOHN W. BODDIE.

SECTION IV.—CUT-WORMS.

From the Southern (La.) Mirror.

In to-day's paper we publish a communication from Col. D. J. Fluker, upon a subject of great interest and importance to southern agriculturists. Col. Fluker is one of the most scientific and experienced planters in the State, and no man is more capable of investigating agricultural subjects than he. His opinions will carry with them great weight and influence; and he will secure the thanks of the community for his assiduous labors in so useful a cause as affording protection to that fragile and delicate but wonderful shrub—the almighty cotton.

East Feliciana, July 3, 1850.

Mr. Editor:—I have learned through the press, and other sources, that the cut-worm has done irreparable injury to the cotton plant this spring, and is still at work on some plantations in the parish. Until this season, I have uniformly been an extreme sufferer whenever they appeared in the country—never escaped before; but, fortunately for me, they have been "few and far between," so far, doing my plants no harm. I think the cotton is now too large for them. It has been my study, for some years, to destroy or escape these worms; finally, for the first time, last year I adopted the plan of burning off cotton and corn stalks, grass, and in fact everything combustible upon the field, in order to furnish as much ashes as possible to the land generally, knowing they are not fond of ashes or lime. This may have been some benefit; but I rely mostly on late ploughing—leaving the cotton land for the last, and breaking it up deep with two horses, just upon planting, say 1st of April; thereby destroying millions of these worms whilst they were generating. By more early breaking up, they can remain under the cotton ridge, and have sufficient time to breed an army before the young plant can possibly grow out of their reach. As a proof of this position, the *few* discovered in my field were of very small size. My cotton crop was planted between the 5th and 15th of April, considerably later than I usually plant. I do not, however, presume that the late planting could have had much to do with it, because the cut-worm is said to be worse upon the replant of May than the older stalks. I leave practical men to draw their own conclusions; still I must cling to mine, that it was the late ridging up of my land which saved me from the cut-worm this year.

If the publication of these hints, hastily thrown together, will have a tendency to relieve the cotton planter, in 1851, from the ravages of this vile enemy of our great staple, I shall be gratified; and you, Mr. Editor, will have done the State some service, for the lot of the Louisiana cotton grower is a hard one, God knows.

Respectfully,

D. J. FLUKER.

SECTION V.—DESTROYING THE COTTON-MOTH.

Mr. Editor:—After having nearly lost three or four crops of cotton by the ravages of the worm, men who heretofore have talked as if they believed that the constitution of this world was something like a system of optionism, or that the farmer had nothing to do but plant, and keep the grass and weeds down, and he had done all that was in his power, now talk on the subject as though they believed that the Creator had bestowed on them faculties to observe and trace cause and effect. They manifest not only a strong spirit toward improving their farms, in making them more productive; but they think they can do something toward checking the advancement of that enemy that has proved so injurious to the South. Men who have been opposed to everything like an interchange of opinion, through the press, in relation to farming, and who have been ready to pronounce everything that is new, either as a humbug or utopian, are now busily engaged in catching flies; and it appears they will have no contentment until the whole fly family is entirely exterminated. And probably it will not be uninteresting to some of your readers, if not profitable, to give you the *modus operandi* how this thing is done; but as regards the effect that it will have, that is a subject on which I am non-committal.

We make a mixture of molasses and vinegar, and put it in plates sufficiently deep to hold the flies after they are caught. Some add a little cobalt; but I don't know that they succeed any better than the others. The plates are placed about over the field, on stakes about the height of the cotton, with boards nailed on their top ends, large enough to set the plates on. The flies are attracted to those plates by the scent of the mixture, and are entrapped. I have, with eighty plates, averaged over 1,000 flies the night, and have taken as many as 45 and 50 from a plate in the morning, that were caught the previous night. I have heard of some persons taking as many as 70 from a plate in the morning, that were caught the night before. There ought to be one plate to each acre of cotton, though I know of no one who has them so thick.

Another way that some are attempting to destroy them, is by striking them down with paddles, their whole force being employed in that way mornings and evenings. If it is a fact that the moth (which some doubt now, I believe,) deposits her egg on the cotton, which makes the worm; then it looks reasonable that, by destroying the flies, the number of worms must be lessened: though I am inclined to believe that the dry weather we have now will be of more advantage in that respect than everything else beside. The flies are very abundant, though I have heard no complaint of the worm. We can't make a large crop in this region, even if we have no worms, for three reasons: it is backward,—the weed is unusually small,—and there are generally bad stands.

I am sorry to inform you that not one of the India cotton seed that I received from the Patent Office, came up; the most of them, I think, were rotten. A few, however, had the appearance of being sound, but were too dry to vegetate.

Respectfully yours,

Sumterville, Ala., August, 1850. J. R. D.

SECTION VI.—THE BOLL-WORM AND "SORE SHIN" IN COTTON.

MESSRS. EDITORS:—There has been a great deal *said* among planters as to means for the destruction of the boll-worm (which is given up to be the greatest evil cotton is heir to), but not much *done*. The means are within the reach of every planter, if he was apprised of it. If the plan that I will lay before you is not effectual, you may take my hat, though a "shocking bad one." In the spring of 1849, my pigs, between thirty and forty in number, ran through into my cotton field. I determined to veto it on seeing any depredations committed by them; but, to my astonishment, they devoured grass, tie-vines, weeds, purslain, &c., and, making diligent search after cut-worms, destroyed them entirely. The result was, they were in high keeping, and few, if any, boll-worms followed. This forced upon me the conclusion, that all worms that prey upon the cotton plant are of the same origin. Every observing mind is aware that the insect tribe is undergoing a constant change—hence, the cut-worm is changed to the moth; the moth deposits its eggs upon the plant, and, by the warmth of the sun, they come forth worms, somewhat different from the cut-worm in form and size. The cotton plant, at this stage, being rather tough for them, they attack the tender forms and pods, hence the appellation boll-worm. I should have tried the experiment farther, but having, since 1849, planted corn in my cotton, I could not allow my pigs that liberty, as there can be no agreement made with pigs not to molest the corn. I am trying the plan this season, and we shall see what we shall see.

Messrs. Editors, if my theory be correct, strike at the fountain head, destroy the parent worm, and you destroy generations to come. Turn in your pigs, and, my word for it, but few worms will be left. But little tutoring will attach them to the field. The mass of planters require a remedy, without money and without price; and, as a dose to a large field, I recommend a large number of pigs—a small field, a small number, &c.

The *sore shin* has been very destructive to our stands of cotton this season; the cause of which, I think, is not generally understood by planters. The *sore shin* is confined to poor land, more particularly to poor sandy hills. My opinion is, that it is occasioned by lice on the leaf of the plant, which runs to a disease. It is admitted by all botanists that the leaf is the lung of the plant, to take up the gases, and prepare the sap to return to the trunk or stem. Too much rain produces lice upon the leaf, which obstructs the laws of nature—the leaf cannot return true sap, consequently the trunk perisheth. In like manner disease the lungs of man, and the trunk will likewise perish. It is confined to poor ridges, because it is slow in growth, and more subject to disease—less vigorous than that on rich land. Lice never produce *sore shin* after the plant obtains a great many leaves. Lice only attack the top and upper limbs; the remaining leaves are ample to support the plant. Some contend that it is the effects of a bruise, or cut with the hoe. If such be the case, why is it confined to hills and poor places? I have often seen wounded stalks occasioned from the hoe. The effect is quite different, and does not come under the head of *sore shin*. I think manuring the places subject to lice, and thorough tillage, will obviate the cause.

I am pleased to see, on the part of some of your correspondents, a disposition to withdraw the firebrand from the camp. I know of many who could, and perhaps would, impart much useful and practical information to the columns of the *Cultivator*, did they not expect to be taken off by the crabbed pens of "crusty" critics.

Yours, truly,

Amite Co., Miss., July, 1853. HEBRON.

SECTION VII.—BIRDS VERSUS INSECTS.

The late Dr. Harris, who was well known for his entomological researches, held the following sentiments respecting birds and insects:

"In order to aid in checking the ravages of noxious insects, protection should be given to their natural enemies. To this end, a stop should be put to the indiscriminate and prevailing slaughter of insect-eating birds and quadrupeds by the murderous gun. Those persons who now waste their time and powder in killing these innocent and useful creatures, would be better employed in planting corn and trees, and in making two blades of grass to grow where only one grew before. Your wood-peckers have already shown themselves to be your friends; let them have all due encouragement."

SECTION VIII.—ANOTHER PLEA FOR THE BIRDS.

THE following interesting passages are from a paper read by Mr. Townsend Glover, before the late meeting of the United States Agricultural Society, and published in the Washington *National Intelligencer:*

"Here, however, let me change the subject, to put in a plea for mischievous birds, which appear to have been sent to keep the 'balance of power' in insect life, which insects would otherwise multiply to such a degree as to be perfectly unbearable, and render the agriculturist's toil entirely useless. A farmer keeps a watch-dog to guard his premises, and cats to kill rats and mice in his granary and barn; yet he suffers an 'unfeathered biped' to tear down his rails, in order to get a chance shot at a robin, wren, or blue-bird, which may be unfortunate enough to be on his premises; and yet these very birds do him more good than either dog or cat, working diligently from morn to dark, and killing and destroying insects injurious to his crops, which, if not thus thinned out, would eventually multiply to such an extent as to leave him scarcely any crop whatsoever.

"Birds are accused of eating cherries and other fruits. True; but the poor birds merely take a tithe of the fruit to pay for the tree, which, but for their unceasing efforts, would otherwise probably have been killed in its infancy. To exemplify the utility of birds, I will give one or two instances that have occurred under my own observation. Some years ago, I took a fancy to keep bees; accordingly, hives were procured, and books read upon the subject. One day a king-bird, or bee-martin was observed to be very busy about the hives, apparently snapping up every straggling bee he could find. Indignant at such a breach of hospitality, as his nest was on the premises, I hastened to the house to procure a gun to shoot the marauder. When I returned, I perceived a grayish bird on the bushy top of a tree, and, thinking it was the robber, I fired, and down dropped a poor, innocent Phœbe bird.

"Hoping to find some consolation to my conscience, for having committed this most foul murder, I inwardly accused the poor little Phœbe of having also killed the bees; and, having determined to ascertain the fact by dissecting the bird, it was opened, when, much to my regret and astonishment, it was found to be full of the striped cucumber bugs, and not one single bee. Here I had killed the very bird that had been working for me the whole season, and perfectly innocent of the crime for which it was sacrificed. After the circumstance, I determined to never let a gun be fired on the premises, excepting on special occasions ; and at present the place is perfectly crowded during spring, summer and autumn, with the feathered songsters, which build their nests even in my very porch, and bring up their young perfectly fearless of mankind; and although cherries, strawberries, &c., do suffer, yet the insects are not a quarter as numerous and troublesome as they were formerly.

"In the southern States, I have seen the bee-martin chase and capture a boll-worm moth, not ten paces from where I stood, and the mocking-bird feeding its nearly grown young on the same insect. Even the ugly toad works for the farmer and gardener, as his food consists of insects more or less injurious. The beautiful and lively green and gray lizards of the southern States, which are seen running on the fence-rail, or amidst the green foliage of trees, shrubs and bushes, and from which they can scarcely be distinguished except when in motion, are ever on the watch for insect prey; and I know of one curious case in which even the mice in the green-house were of service, for they had rooted up the earth round several potted peach trees, in order to devour the chrysalis of the peach-tree borer."

SECTION IX.—RED RUST AND BROWN RUST.

From the Columbia (S. C.) Planter.

DEAR SIR:—I feel ashamed of not having yet complied with your request, that I should send you a treatise on the manufacture, application and effect of manure. I will, however, compromise with my conscience, by promising to do so in a week or two. The fact is, I am so conscious of devoting less attention and labor to that department of plantation economy, than its importance demands, that I feel a repugnance to seeing my deficiency formally and mathematically computed. My present impulse is to discourse on the inexplicable subject of rust in cotton, and I will not thwart it.

In this section of country, we have two species of rust—the red or common rust, and the brown or French. I cannot give you the derivation of the latter term, but it is of general prevalence in this neighborhood. The red rust is that to which all varieties of land in this district are more or less liable; and the brown rust, or French, that which is only found on black-jack soils, and on the flat lands of the description of those on Dutchman's and Wateree creeks.

As Humbug jr. affirms, many theories have been adduced to account for the origin of the red rust, the partizans of each believing firmly in his own favorite, and denouncing those of others; and he accordingly treats as an absurdity, a creed of mine, which I consider I have incontestibly proven by an experience of nine years.

I commenced to till my present lands, and at the same time to cultivate the short staple-cotton, nine years ago, under the tutorage of an observant, experienced, and skilful overseer. In the early part of the summer, he remarked to me, that we should have to keep our cotton fields free of poke-weeds and briars, if we meant to escape rust. This being a new idea to me, I of course ridiculed it, and so unmercifully too, that, as he afterwards told me, he forthwith determined that I should purchase belief by expensive experience. Accordingly, towards fall, he carried me to three several spots of rust, in as many different fields, which he had contrived to produce by leaving poke-stalks in or around stumps, which happened to be there located. They were the only spots of rust I had in my crop, and from every other portion of it, had the poke been carefully eradicated. This coincidence staggered me, and its repetition for nine consecutive years, has confirmed my faith.

In riding by the fields of my neighbors, I have seen poke-stalks suffered to grow among the cotton, and have predicted to a companion (correctly, as it was proven,) that rust would be the consequence. On the other hand, I have first seen rust, and on searching for it, have found poke.

I do not say that rust may not originate without the presence of poke; nor do I believe that, like the celebrated upas tree, it exudes poison, deleterious to surrounding vegetation; but simply that poke, briars, strawberries, and perhaps other plants, are more liable to the disease than cotton; and having first become affected, communicate the disorder to their more healthy neighbor. Where poke has been repeatedly cut down in the early part of the season, and is suffered to grow at a late date, I believe it is harmless; for it does not appear to be liable to the disease till it has reached maturity, and commenced to decline.

A friend of mine spent a part of the last summer in this vicinity, and occasionally delighted me with a visit. At our first interview, he began, in unmeasured terms, to denounce the folly and superstition of "some ignorant citizens of the district," in believing that poke could produce rust. Like St. Peter, I was at first ashamed to confess my faith in so despised a doctrine, and, if I did not deny, certainly did not avow that I was a disciple. At each subsequent interview, I was gratified to observe that the opposition of my friend was melting away under the influence of accumulating proof, till at length, when I rallied and came to the rescue, his offensive warfare degenerated into mere defence of his doubts, and he finally determined to risk his remaining strength upon the issue of a single experiment, an opportunity for which then presented itself. In the midst of a large, healthy, flourishing field of cotton, he saw a small spot of rust, and he determined to surrender at discretion, if in the centre he should find pokeweed. The poke was found, and he acknowledged himself a convert.

I may properly close what I have said of red rust, by stating, as a corollary, that though we do not know what produces it, in poke or cotton, nevertheless, if poke, briars, &c., are more liable to the disease than cotton, and can communicate it, it is wise not to suffer them to take root in our fields. We know not the cause of the origin of yellow fever or smallpox, yet we know that they are communicable and infectious, and avoid persons and places suspected of being tainted with their influence.

Of the French or brown rust, though I suffer from its effects, I have very little to say. Some people have attributed it to the presence of iron in the soil, in some of its chemical forms. Others (and I am among them), believe that it is caused by an undue proportion of lime in the soil, causing the plant to scald under the superadded influence of heat and water. I once saw the late Dr. James Davis, of Columbia, analyze soil taken indiscriminately from the land that is liable to the French, and the result was seven and a-half per cent. of carb. of lime. Now, this is certainly a much stronger proportion than even three hundred bushels of rich marl would give uniformly to the whole mass of an acre of soil, if thoroughly amalgamated with it. The flat creek lands, upon which this disease prevails, are not the alluvial bottoms; these are of a distinct character from the former, which lie between them and the sandy ridge.

I have found that late ploughing promotes the French, and that compost manure is the best preventive.

Yours, very respectfully,

Fairfield, August 30, 1843. FARMINGTON.

SECTION X.—"BLUE COTTON."

GENTLEMEN:—I have been a subscriber to the *Cultivator* for the past year, and have just forwarded to you, through the Postmaster, the amount of subscription for another copy, the receipt of which you will acknowledge by sending me the first number for 1844.

How does it happen that I have never, to my recollection, seen in your paper a single paragraph in relation to Sea Island cotton? Can it be that you have so few patrons on the seaboard, or that they send you no communications? I have waited patiently myself for such, and perhaps others have done the same, from the same motive. Our lands yield in value a large portion of our exports, and it is a matter of considerable interest, to our small number at least, to give and receive information on that subject; and, although I am by no means an adept in the culture of the long staple, I *might* occasionally throw out a hint which might be extended and improved upon by others more capable than myself. But to the object of this communication: We have large portions of land on the main, adjoining our salt-water rivers and inlets, such as live oak flats, &c., which produce what is termed, in our fraternity, Blue Cotton, from, I presume, the blueish cast of the plant. These lands are very rich, and produce fine crops of corn, but so far as I am acquainted, there has been no remedy applied for Blue Cotton, which they almost invariably produce. By this term we mean such cotton as comes up and grows very luxuriantly, without any fruit, reaching at times the height of eight or ten feet, having large leaves, with crimped edges, and of a deep lead color, so much so that a spot in the field may be recognised as far as the plant can be distinguished. At other times, depending perhaps on a very wet season, the plant, after growing several feet, and bearing well, sheds all its fruit and becomes *blue.*

This is a serious difficulty with our strong lands, and I hope among your many readers, some one may be able to suggest a remedy for the evil. It has been generally supposed among us, that land containing a large quantity of iron would have this effect—why, I know not; but if such is the fact, it appears to me that lime would be a good application, and it is my intention to try it. The chemical action of lime on the organic substances of which our strong low flats and swamps contain a great deal, is very considerable, and this is not only in reference to vegetable remains, but it acts with equal energy upon the dead and living animal matter. Its operation, therefore, *may* effect a change in the production of the plant. Besides, if the soil contains sulphate of iron, this is decomposed by the lime, which, uniting with its sulphuric acid, forms the sulphate of lime, which is commonly called gypsum, and which is universally admitted to be a great fertilizer of the soil. Now the question is, will this chemical process have the effect of changing those matters in the soil which cause our cotton to turn *blue*. Experience, of course, will be our surest guide on this subject, but it would be deeply interesting to read the views of some of your learned correspondents in relation to it.

Respectfully, &c.,

Bryan County, January 1, 1844. AGRICOLA.

P. S. I have been using, for some time, the plough in the cultivation of the Sea Island cotton, with advantage, and I intend, this year, further to facilitate my work by the side-harrow and the cultivator. Be good enough to say, in your next number, if I can obtain the latter implement in Augusta, whose make, at what cost, and whether they will answer between beds four and a-half feet apart. They ought to be made so as to be moved for a greater or less distance.

SECTION XI.—THE DRY ROT IN COTTON

EDITORS SOUTHERN CULTIVATOR:—Permit me, in behalf of myself and neighbors, to make known the existence of a disease in our cotton to a much greater extent than ever before known by us, called the "dry rot," and ask you some questions as to the probable cause.

The disease we speak of attacks the top bolls. The seed and lint first rot and turn black; then a sore or scab appears, resembling a puncture with a sharp instrument. This extends quite over the surface of the boll, and very frequently—after the disease has taken possession of the whole pod—it opens its prongs and presents a thoroughly rotten state in all its parts. So far as the writer's observation extends, it is most injurious to sandy soils, and on these it appears most malignant in those fields which have been longest in cultivation. It is seen, however, in places where the lands have been well manured and cultivated. I am informed the lime or cane brake lands are suffering to some extent with it.

In this vicinity, it is felt as a serious drawback on our crops from one hundred to three hundred pounds per acre, all of our crops will suffer from it, and this after the bolls seemed to have matured. Hence, we have conversed on the subject a good deal, and the writer has concluded to call your attention to it, and ask, if the cause is known to yourself or older planters than we are. The specific inquiry or inquiries we would ask of your better judgment, are, whether this rot is probably caused by peculiarity of soil? Or, is it the result of the seasons (these have of late been uniform)? Or the mode of cultivation; has this influenced it? Is there any reason to credit the conjecture that one variety of cotton is more liable than another to this disease? This year the rot is doing so much harm to our cotton as to call for examination and remedy, and if it should increase its ravages from year to year, it would be felt as a serious evil.

I hope the readers of the *Cultivator*, as well as yourselves, will give attention to the subject—all planters are interested, at least in one point, and that is as the extent of the injury.

Very respectfully, I subscribe myself.

Alabama, October, 1855. BEAVER BEND.

If the cotton plant should suffer as much from premature decay, in the course of a few years, as the potato plant has, the occurrence will not surprise us. Gangrene, whether "dry" or otherwise, in vegetable and animal tissues, arises commonly either from the weakening of vital force by improper nourishment, the presence of a poisonous substance, or from some unknown constitutional defect. The source of "canker," which attacks fruits and fruit trees,—of the potato rot, and the rot in the seed and lint of cotton,—is involved in great obscurity. Whatever may be the primary exciting cause either of the premature extinction of life in the parts affected, and of their rapid dissolution, the warmth and humidity of the surrounding atmosphere may be such as to favor the destructive increase of the malady. All living beings are creatures of circumstances, with many of which the wisest are yet unacquainted. We know that inflamed flesh is apt to mortify; and that the dead limbs of a man, like those of a tree, may even drop off by purely vital and chemical processes. Nature has many secrets in vegetable and animal life and death, that human science may never penetrate nor reveal. But this fact should not prevent our studying all the phenomena of vitality, as displayed in ultivated plants and domesticated animals. English farmers find it impossible to grow common red clover on land where this plant has flourished for a half century, without being able to assign any good reason for the fact that the soil is now "clover sick." *A change of crops*, in all such cases, has been the best remedy, where others failed. The fastest horses, with the largest constitutional resources, may be broken down by over work; and why may not the vital resources of the cotton plant be over tasked by those who seem willing to drive cotton culture as one drives a free horse, till he falls dead and *rots?*

The over-feeding of an animal is a poor remedy for pushing him beyond his natural powers of endurance; and, by a parity of reasoning, to surcharge the vessels and cells of a plant with liquid manure, is not a proper preventive of "rot," in its sound or diseased system. Potatoes rot most, when thus treated.

According to our ideas, diseases in vital organs and functions are seldom viewed so philosophically as the present advanced state of physiological science renders practicable. If we were to say that the earth and climates, including air and water, produce murrain in cattle far more in some localities than in others, as similar elements of disease produce bilious affections in the human family, the true sources of these well-known maladies would be but poorly explained. Unquestionably, many causes often coöperate to weaken the vital principle in plants and animals; and the early death and dissolution of a single cell in the fruit of a cotton plant, are doubtless sufficient to bring on the chemical disorganization of the whole boll, if not of the whole plant. The rotting or decay of every tissue is purely a chemical process, however this disorganizing operation may have originated.

If we have read the agricultural literature of civilized nations aright, such diseases as the blight on pear trees, the premature rotting of apples, potatoes, and other vegetables, and the rot in cotton, are not likely to diminish in the aggregate, until farmers know more of the laws of nature, and of the true principles of farm economy, than they now do. They cannot systematically obey laws of which they know little or nothing. So long as farmers in Western New York raised potatoes on fresh virgin soils, they were exempt from the potato rot as a prevailing distemper; and we fear that, as cotton is cultivated year after year on the gradually deteriorated lands of the South, there is no strength of vitality in this weed to protect it, *indefinitely*, from constitutional deterioration, and its natural consequences.

In an excellent article on "Cotton culture, and selection of Seed," in our last issue, Mr. A. W. Washburn, of Yazoo, Miss., says that his crop averages a bale of cotton of 400 lbs. to the acre, although he plants "on prairie land twenty-five years under hard cultivation, without manure." He makes ten bales to the hand, and probably is not at all injured by the rot. Such facts speak well for the natural resources of his soil; but "hard cultivation" for twenty-five or thirty years more without manure, may so change the physical and chemical properties of this land as to weaken the cotton plants which grow therein; and their seed, planted in districts where the rot prevails, will yield crops equally subject to the malady. If our cotton rotted, as described by "Beaver Bend," in our last number, we should grow, or obtain from another, seed produced on fresh land to plant hereafter. The land on which cotton is to be cultivated, should be ploughed an inch or two deeper than usual, to give the growing plants the benefit of a better pasture in *fresh earth.* The subsoil is often full of virgin, fertilizing resources, which superficial, shallow tillage never reaches. There is a striking analogy between the healthy pasturage of domesticated plants and domesticated animals. If unchanged into new and fresh pastures, cattle soon eat down, and finally kill out those nutritious grasses and herbs best adapted to form pure blood, sound flesh, nerves and bones. They may still subsist, and propagate their kind for several generations; but under far less favorable circumstances, and more subject to casualities. The over-cropping of their land is a similar folly. It parts with some element of vegetable nutrition, unseen and unappreciated by the cultivator and there is left to him a disordered soil, yielding cotton plants of unnatural, unsound growth, which Nature disowns, vitality deserts, and chemical laws speedily resolve into their original elements. As every thing that lives, decays or "rots" sooner or later, it is a question of time and circumstance, *when* and *how* this final result shall be attained. A reasonable supply of potash in the soil is known to promote the healthy growth of the woody fibre in plants, (which forms the lint of cotton, and a part of its seed,) and also favors the perfect organization of starch, sugar, oil, and the so-called protean compounds; therefore let wood ashes be applied to the "sandy land" where cotton rots.

It is an unwise, a bad system of cultivation, that makes so many old and deserted fields in the cotton-growing States. Nature never gets tired of growing crops of forest trees, even on the very poorest lands of the South. This fact is full of instruction. Man wantonly violates her laws, and disease, in a thousand forms, is sent to chastise him into better conduct. How far Providence will punish the impoverishment of arable lands, we have all yet to learn. It will, however, be sufficient to compel a reform in our present system of tillage and husbandry. If one degree of rot, of "murrain," or other calamity, is insufficient to bring us back to the straight and narrow path of agricultural duty, another, and still another degree of chastisement will be added, until, penitent and willing to obey the laws of his Creator, man will properly feed the land that both feeds and clothes him. L.

SECTION XI.—ROT IN COTTON.

We copy the following from the *Liberty* (Miss.) *Advocate*, of a recent date :

MR. FORSYTHE :—In a former communication, I alluded to the rot in cotton, which when properly considered, deserves more than a passing notice. Millions of inhabitants are dependent upon the culture and manufacture of the great southern staple for employment. The disease does not affect the northern producers, when they can obtain enormous prices for their produce, or the manufacturer, when they can buy our staple for a mere song—which is not warranted by the proper statistics. In 1849, the manufacturing companies of Manchester adopted the following resolution:

"*Resolved,* That we consider all reports relative to a short crop, such as overflows, worms, &c., altogether humbug." They are now raising the cry of large crops, in order to keep down the prices. I fear we have given them too much rope to take it up with ease, unless we make a long pull, a strong pull, and a pull altogether. They are now like the negro's horse—have two good eyes, and won't see. It is fearful to contemplate, when we consider the ravages the disease is making upon the cotton plant, and so little said or done to remedy the evil. Let us take a retrospect of the cotton plant. When the Deity in his goodness gave our forefathers the virgin soil—with their rude agricultural implements—the old black seed to cultivate—which was so peculiarly adapted to his situation—when cotton gins were among the things uncumbered, the pioneer, with his wife and his little ones, after their daily toil in gathering their cotton in, assembled around their pine-knot fire to disengage the seed from its linty fibre, whilst the mother and daughters converted it into yarn, to barter it for the necessaries of life. Thus time passed on, until the plant began to run out, and the rot made its appearance, which caused a new importation of seed, different from the black seed; which caused arts and sciences to be put in motion to prepare it for market—then the disease disappeared.

Time passed on, and no very great change was made in the seed, until Col. Abby, of Mastodon notoriety, made a fortune in a very short time by selling his seed at fifteen to twenty, and even fifty dollars per bushel, which went off like hot cakes—proved to be more valuable than the cotton itself. There was a general stampede among the planters who should make the next fortune selling seed, and at the same time heaping denunciations upon the Colonel. By a little care in selecting, and an improvement in name, we soon had a catalogue of names, such as Alvarado, Brown, Pitt, Willow, Hogan, Sugar Loaf, Silk, Vick's 100-seed, and a host of others. Those that succeeded best in giving a big name, and puffing most, bore off the palm, and wore the title of Colonel, and even went farther—for instance, Gen. Mitchell's Prolific Pomegranate, &c. Each of the above varieties succeeded very well for two or three years, and then sank below par. All the notoriety was given to the variety, when half was entitled to a change of latitude. The different varieties have been mixed up so, until it has become corrupted, and the corruption has become epidemic.

We are now where our forefathers were with the old black seed. Many of our old standard planters pronounced it identically the old black rot. I have been a close observer of the disease since it made its appearance. It seems to be worse when we have a warm, cloudy spell of weather, of five or six days. Apparently all the bolls will mildew and rot in a few days. One would suppose that it was atmospheric. Not so. We had just such weather ten years ago, when the rot was not known.

I see a certain M.D. has sent some beetles and diseased bolls to the Smithsonian Institution for examination. I cannot reconcile myself that it is the effect of insects. If it was, we would have had an immense quantity of it during the reign of the army-worm in 1846, and the boll-worm since. They stripped the foliage, cut the rind from the boll, punctured the pods, and even embedded themselves into them, yet they opened beautifully, and the disease was not known.

A correspondent of the *American Cotton Planter* says it is caused from the want of new, healthy, and sufficient quantity of pabulum. I must beg to differ with him. Some of our old hills have suffered these many long years for a want of a sufficient quantity of healthy food, yet the rot did not appear. If any soil shall produce healthy pabulum, it should be fresh land. My experience is, the disease is a little worse on fresh land than old land. I cannot see why the disease has not made its appearance before now, if this theory is true. Many planters say, a change of seed a short distance is beneficial. In this I agree; but I prefer a change from a northern latitude, and a district that is not infected, for the following reason: The best latitude for cotton has been considered that of Vicksburg. The latitude has gradually been going further north. The crop of our country from 1840 to 1850, excelled, per acre, that of Yazoo or Holmes county. Since 1850, up to the present time, we have retrograded, while they have increased —the boll-worm has disappeared, and the rot is scarcely known. The picture is truly discouraging. We may obey the mandates of the Scriptures, "What thou doest, do with all thy might," and yet be but little better off than "One that provideth not for his household."

NEBRASKA.

CHAPTER V.

ANALYSES OF THE COTTON PLANT, WITH SUGGESTIONS AS TO MANURES, ETC.

SEC. I.—SHEPARD'S ANALYSIS OF COTTON WOOL AND SEED.

1. *Cotton Wool.*

ONE hundred parts weight of cotton wool, on being heated in a platina crucible, so long as brightly burning gas continued to be emitted, lost 86·09 parts—the residuum being a perfectly charred cotton, which, on being ignited under a muffle, until every particle of carbon was consumed, lost 12·985, and left almost a purely white ash, whose weight was rather under one per cent., or 0·9247. Of this ash, about forty-four per cent. was found to be soluble in water. It contained 12·88 per cent. sand, which must have been acquired adventitiously, in the of silicious process of harvesting the fibre. Deducting the sand from the ash, the constitution of the latter is as follows:

Carbonate of Potassa (with possible traces of Soda), .	44·19
Phosphate of Lime, with traces of Magnesia, .	25·44
Carbonate of Lime, ...	8·87
Carbonate of Magnesia, ...	6·85
Silica, ...	4·12
Alumina (probably accidental), ...	1·40
Sulphate of Potassa, ...	2·70
Chloride of Potassium, Chloride of Magnesium, Sulphate of Lime, Phosphate of Potassa, Oxide Iron, in minute traces, } and loss, . .	6·43
	100·00

But, since it is obvious, that the carbonic acid in the above-mentioned salts must have been derived during the incineration of the cotton, the following view will more certainly express the important mineral ingredients abstracted by the cotton from the soil, for every 100 parts of its ash:

Potassa (with possible traces of Soda),	31·09
Lime,	17·05
Magnesia,	3·26
Phosphoric Acid,	12·30
Sulphuric Acid,	1·22
	64·92

For every 10,000 lbs. of cotton wool, then, about sixty lbs. of the above-mentioned ingredients are subtracted from the soil, in the proportion indicated by the numbers appended, *i. e.*, (omitting fractions:)

Potassa,	31 lbs.
Lime,	17 "
Magnesia,	3 "
Phosphoric Acid,	12 "
Sulphuric Acid,	1 "

Several queries were submitted to me, along with the sample to be analyzed, relative to the effect of soils on cotton. I regret to state that the almost total ignorance in which we are still left, respecting the composition of the varieties of this fibre, and the soils producing them, prevents me from hazarding any explanations on the subject. This is the first destructive analysis ever made (at least so far as my knowledge extends) of the cotton wool. Nor am I acquainted with the properties of the soil which afforded it. Prior to any dedúctions, it is clear we must know the composition of each variety of cotton, as well as that of the soil it affects. At present, I can only venture on connecting together two facts, which appear to occupy important relations to one another. The soil of St. Stephens, which is said by F. A. Porcher, Esq., to be a stiff, clayey loam, produces the strongest and finest fibre of the Santee varieties. The Sea Island qualities are supposed to owe their superiority to the use of marsh-mud, which I have ascertained to be a clayey admixture, rich in alkalies and alkaline earths. Whether the similarity between these two staples is influenced most (if it is affected at all), by the chemical or mechanical qualities of the soils producing them, it is impossible to decide. It is also conceivable, that the two sets of qualities may conspire to one and the same end.

2. *Cotton Seed.*

One hundred parts, heated as above, lost 77·475, and the thoroughly charred residuum, burned under the muffle, left 3·856 parts of a perfectly white ash. The composition was found to be as follows:

Phosphate of Lime (with traces of Magnesia),	61·64
Phosphate of Potassa (with traces of Soda),	31·51
Sulphate of Potassa,	2·55
Silica,	1·74
Carbonate of Lime,	·41
Carbonate of Magnesia,	·26
Carbonate of Potassium,	·25
Carbonate of Potassa, Sulphate of Lime, Sulphate of Magnesia, Alumina, and oxides of iron and manganese, in traces, and loss,	1·64
	100·00

In comparing the above table with that afforded by the cotton wool, a marked dissimilarity presents itself. The ash of the cotton seed is four-fold that of the fibre; while the former has also treble the phosphoric acid possessed by the latter, as will the more clearly appear, when we present the analysis under another form, corresponding with the second table under cotton wool:

Phosphoric Acid,	45·85
Lime,	29·79
Potassa,	19·40
Sulphuric Acid,	1·16
	95·70

From the foregoing analysis, it would appear difficult to imagine a vegetable compound better adapted for fertilizing land, than the cotton seed; nor can we any longer be surprised at the well known fact, that soils long cropped with this staple, without a return to them of the inorganic matters withdrawn in the seed, become completely exhausted and unproductive.

SECTION II.—ANALYSES OF THE COTTON PLANT AND SEED: WITH SUGGESTIONS AS TO MANURES, ETC.

THE natural history of the cotton plant,* and improvement in its culture, in the cotton-growing States, are interesting

* German, *Kattonwolle, Baumwolle;* Dutch, *Ketoen, Boomwol;* Danish, *Bomald;* Swedish, *Bomull;* Italian, *Cottone, Bombagia;* Spanish, *Algodon;* Portuguese, *Algodno, Algodeiro;* Russian, *Chlobts-chataza bumaga;* Polish, *Bawelna;* Georgian, *Bomba, Bamby;* Latin, *Gossypium;* Greek, *Bombyx, Ylon;* Mongul, *Kobung;* Hindoo, *Ruhi;* Malay, *Kapas;* Indian, *Kopa;* Chinese, *Cay-Haung, Hoa-Mien.* Skinner, the Etymologist, says,

subjects. Originally the production of the tropics, it has, in our country, travelled far into the temperate region, and flourishes on a belt of several hundred miles wide; extending from Virginia along the sea coast to our western limits on the Gulf of Mexico. Congeniality of climate, seasons and soils, has carried the cultivation of this plant, which is not certainly ascertained to have been indigenous to the United States, much further than it was at first expected it would ever be extended; and it has become the staple of all those parts not actually mountainous in the southern States. Whilst its culture has most rapidly advanced and increased in every section, the planters of the old cotton-growing States, from the exhaustion of their soils, and the lack of proper systems of rotation and manuring, have been thrown in the back ground in the scale of profitable production by their more favored rivals, the fortunate possessors of the virgin lands of the south-west. If this deficiency is ever to be remedied—if the fertility of those soils, worn out in the oft-repeated production of cotton, is ever to be restored, and permanently improved for the future culture of this crop, or for other systems of tillage, it must be done under a proper understanding of what constituents are to be restored to the soil, to supply the places of those of which it has been robbed. How far a correct analysis of the cotton plant and seed, will enable the present generation of planters to remedy the lack of fertility in their impoverished soils, and enhance their future productiveness for this crop, it is difficult to determine; but it is no

that cotton is so called from its similitude to the down which adhered to the quince, *malie cydoniis*, which the Italians call *cotogni*, and *cotoquy* manifestly *a cydonis*.

Gossypium, or cotton, a genus of the polyandria order, belonging to the monadelphia class of plants, and in the natural method of ranking, under the thirty-seventh order, Columniferæ —*Encyc. Britannia, vol.* 8, *p.* 21.

matter of speculation, to assert, that it is essentially necessary for the soil, for the cereal crops, that the past industrious despoiling of the natural elements, should furnish a guide for their restoration. The analytical investigations made by the author, and for their correctness receiving the sanction of Professor Von Liebig, the most celebrated chemist of the age, and given to the world in their present shape, are not intended as the basis of a new theory for the production of the cotton plant, but merely as suggestive of aids, and by returning to the soil what has been taken from it, bring about a restoration of fertility, which will render its cultivation profitable to agriculturists in any other marketable crops. When, however, we reflect that of the one thousand millions of pounds of cotton, produced in the world, upwards of five hundred and fifty millions of pounds are grown in the United States, we readily see that the importance of this crop—swelling to this enormous amount since 1784, when it was doubted at Liverpool, that so much as *eight bales* could be produced in this country—demands all those scientific aids by which other nations have fostered their staple agricultural productions, and thereby contributed to national prosperity. England, by her commercial enterprise, assumed the pinnacle of national rank. The cotton plant, its productions and adaptation to human wants, by manufacturing skill, will give the blood to invigorate our prosperity. What a picture of prosperity would be presented, if we manufactured in South Carolina all the cotton grown in the State, and had sufficient commercial capital and enterprise to concentrate the exportation and exchange of the manufactured material at our queen city, Charleston! Added to this, how much more pleasant would be the prospect ahead, if the cultivation of this crop was so regulated and carried out that it would fit the soil for the increased after-production of the grain crops—those crops so essential to the prosperity of the world!

Analysis of the Ash of the Cotton Plant.

Qualitative Analysis.—A part of the ash was taken and boiled with distilled water, then filtered, the filtrate acidulated with nitric acid, and then treated with nitrate of silver, (AgO, NO_5.) A white precipitate of chloride of silver was formed, showing the presence of chlorine.

On adding muriatic acid to another part of the ash, an effervescence took place, showing the presence of carbonic acid.

Another part of the ash was taken and dissolved in muriatic acid, and evaporated to dryness; then moistened with muriatic acid, and digested with water—a residue consisting of coal, sand and silica, remained insoluble. The presence of silicic acid was proved, by boiling the residue with potassa, (free of silicic acid) and evaporating the filtrate in the presence of muriatic acid, to dryness, then moistening with muriatic acid, and dissolving with water, the silicic acid which remained insoluble. A portion of the liquid, freed from sand, coal, and silicic acid, was nearly neutralized with ammonia, when, upon the addition of acetate of soda, a white precipitate of phosphate of iron was formed.

To a part of the liquid filtered from this precipitate, ammonia was added, which formed a white precipitate, showing that all the phosphoric acid was not in combination with iron.

To another part of the liquid filtered from the precipitate of phosphate of iron, oxalate of ammonia was added, which formed a white precipitate of oxalate of lime.

The liquid filtered from this precipitate gave, on the addition of phosphate of soda and ammonia, a precipitate of phosphate of magnesia and ammonia, showing the presence of magnesia.

Another part of the liquid, freed from sand, coal, and silicic acid, was boiled with an excess of baryta water, and filtered. The excess of barytes in the filtrate was removed by carbonate of ammonia and ammonia, and filtered—the filtrate was evaporated to dryness, and dissolved in a small quantity of water. A part of this solution was treated with bi-chloride of platinum; a yellow crystalline precipitate was formed, showing the presence of potassa.

A part of the residue was tested with the blow-pipe for soda; the presence of which was proved.

A portion of the liquid, freed from sand and silica, was treated with chloride of barium; a white precipitate of sulphate of barytes was formed, showing the presence of sulphuric acid.

Quantitative Analysis.—6·181 grammes of the ash was digested with muriatic acid, and evaporated over a water bath to dryness. The residue was gently ignited, and moistened with muriatic acid, then let stand for half an hour, after which it was digested with water, and filtered upon a weighed filter. The coal, sand, &c., remained upon the filter, and was washed out with boiling water, until, on evaporating a drop of the filtrate on the platina foil, no residue remained.

The filter was now dried, and all the sand, coal, &c., carefully separated, (in order not to damage the filter), after which, the substance which was on the filter was boiled with potassa in a platina basin over a water bath for one hour; then filtered upon the same filter, washed out with distilled water, and dried at two hundred and twelve degrees, until it remained at a constant weight. After deducting the weight of the filter, there remained 0·621 grammes of sand and coal.

The part soluble in potassa, was mixed with muriatic acid, (HCl,) and evaporated over a water bath to dryness; then ignited, and moistened with muriatic acid, (HCl,) and dissolved in water, filtered and washed, then dried and burned. It weighed, after burning, 0·403 grammes, silicic acid, (SiO_2.)

The liquid filtered from the sand and silicic acid, (measured in a graduated tube,) was found to contain four hundred and eighty cubic centimetres, which was divided into three equal parts of one hundred and sixty cubic centimetres, each = 2·060 grammes of the ash, for each one hundred and sixty cubic centimetres of the liquid.

These three parts will be termed A, B and C.

In A, the phosphate of iron, lime and magnesia, were estimated.

In B, the sulphuric acid, and the entire quantity of phosphoric acid.

In C, the alkalies.

A.

The liquid A was nearly neutralized with ammonia, then acetate of soda and free acetic acid were added. The precipitate was left standing for twenty-four hours, after which it was filtered, and washed out with boiling water, then dried and burned. It weighed 0·346 grammes, or, for the entire liquid, 1·038 grammes of $2Fe_2O_3 3PO_5$, or 0·507 grammes of Fe_2O_3 (oxide of iron).

The liquid filtered from the precipitate of phosphate of iron was treated with oxalate of ammonia. The precipitate of oxalate of lime, was filtered, washed, dried and burned. It weighed, after burning, 0·643 grammes of carbonate of lime, (CaO,CO_2,) or for the entire liquid, 1·929 grammes of CaO,Co_2, = 1·092 grammes lime (CaO).

The liquid filtered from the oxalate of lime, was evaporated over a water bath, to a smaller volume, then phosphate of soda and ammonia were added, and the precipitate left standing for two days, after which it was filtered and washed out with water containing one-eighth of ammonia, and burned until it was white. It gave 0·301 grammes of $2MgO, PO_5$ (pyro-phosphate of magnesia), or for the entire liquid, 0·903 grammes = 0·330 grammes MgO (magnesia).

B.

The solution B was precipitated while boiling with chloride of barium, and left standing on a sand-bath for twenty-four hours, then filtered and washed with boiling water, dried, and burned. It gave 0·079 grammes sulphate of barytes (BaO, SO_3), or for the entire liquid, 0·237 grammes (BaO, SO_3) = 0·081 grammes sulphuric acid (SO_3).

The liquid filtered from the precipitate of BaO, SO_3, was mixed with per-chloride of iron and acetate of soda, and boiled for five minutes in a large flask; then the precipitate of phosphate of iron, and basic acetate of iron, was filtered while warm, and washed with boiling water, until on evaporating a drop of the filtrate there remained no residue.

The precipitate was dissolved while moist, in as small a quantity of muriatic acid as possible. Tartaric acid and ammonia were now added in excess, when to the clear yellow-colored solution, a mixture of sulphate of magnesia and chloride of ammonia was added, to prevent a precipitate of magnesia. The precipitate was left standing for two days, after which it was filtered and washed out with water containing ammonia. When dried, burned and weighed, it gave 0·442 grammes of $2MgO, PO_5$, or for the entire liquid, 1·326 grammes of $2MgO, PO_5$ = 0·837 grammes phosphoric acid (PO_5).

C.

Baryta water was added to this solution, until an alkaline

reaction had taken place, then boiled and filtered. The excess of barytes in the filtrate was removed with carbonate of ammonia and free ammonia—the filtrate was evaporated over a water-bath to dryness, and ignited until it was free from all ammoniacal salts, then dissolved in water. Some magnesia remaining insoluble, was filtered off, and the filtrate again evaporated to dryness, and ignited, then weighed. It gave 0·770 grammes of the chlorides of the alkalies, which is for the entire liquid, 2·310 grammes. These alkalies were again dissolved in a small quantity of water, and the potassa estimated with bi-chloride of *platinum*, which gave, after being evaporated with alcohol over a water-bath, 2·356 grammes of double chloride of potassium and chloride of platinum (KCl, $PtCL_2$), or for the entire liquid, 7·068 grammes (KCl, $PtCl_2$). This represents 2·157 grammes chloride of potassium (KCl), or 1·326 grammes potassa (KO).

There remains, consequently, after subtracting the chloride of potassium from the chlorides of the alkalies, as follows, the amount of chloride of sodium, which is estimated as loss, thus 2·310 KCl, NaCl—2·157, KCl. =0·53 (NaCl), chloride ot sodium. 2·970 grammes of the ash was boiled with distilled water, and filtered. The filtrate was acidulated with nitric acid, then precipitated with nitrate of silver. It gave 0·044 grammes of chloride of silver ($AgCl$), or 0·022 grammes chlorine (Cl), also, 0·153 grammes, NaCl,—0·037 grammes, NaCl. = 0·116 grammes chloride of sodium ($NaCl$), = 0·061 grammes soda (NaO).

1·066 grammes of the ash gave 0·168 grammes carbonic acid (CO_2). The following is the per centage of the constituents in 100 parts of the ash.

GRAMMES FOUND.		PER CENTAGE.	
Silicic Acid, . . .	0·403	6·50	SiO_2.
Sand and Coal, . .	0·621	10·04	Sand and Coal.
Oxide of Iron, . .	0·507	8·20	Fe_2O_3.
Oxide of Lime, . .	1·092	17·66	CaO.
Oxide of Magnesia, .	0·330	5·33	MgO.
Sulphuric Acid, . .	0·081	1·31	SO_3.
Phosphoric Acid, . .	0·837	13·37	PO_5.
Potassa, . . .	1·362	22·01	KO.
Soda, ...	0·061	0·99	NaO.
Chloride of Sodium, .	0·307	0·05	NaCl.
Carbonic Acid, . .	0·168	15·72	CO_2.
		101.19	

ANALYSIS OF THE ASH OF COTTON SEED.

Preparation of the Ash.—The seed were burned in a Hessian Crucible, with a muffle. Only a slight red heat was necessary to burn them perfectly white.

For estimating the amount of water, 6·406 grammes of the seed were taken and dried, at 212 deg., until they remained at a constant weight. They gave 0·646 grammes water, = 10·08 per cent.—in 100 parts of *the seed.*

Estimation of the Ash.—The seed were dried, until they remained at a constant weight, then burned in a platina crucible. 6·587 grammes of the dried seed gave 0·237 grammes ash—equal 3·8 per cent. ash, in 100 parts of the dried seed.

The qualitative analysis showed that all the constituents were present, which were in the ash of the plant, with the exception of carbonic acid.

The quantitative analysis was carried out similar to that of the ash of the plant heretofore described.

The following are the results—1·882 grammes of the ash was used:

FOUND.		PER CENTAGE.
Phosphoric Acid,	0·667	35·43 PO_5.
Oxide of Iron,	0·075	3·33 Fe_2O_3.
Coal,	0·020	1·05 Coal.
Sulphuric Acid,	0·060	3·19 SO_3.
Oxide of Lime,	0·204	10·88 CaO.
Oxide of Magnesia,	0·200	10·61 MgO.
Potassa,	0·523	27·82 KO.
Soda,	0·051	2·75 NaO.
Silicic Acid,		Trace.
Loss and Chlorine,		4·84
		100·00

Suggestive Remarks.

On examining the foregoing analysis of the cotton seed, we see that they abound in the phosphates and alkalies. Drs. Will and Fresenius, in their analysis of the cereal grains, show that wheat also abounds largely in these constituents.

In order to enable the reader make the comparison, we give the analysis of red and white wheat, as published by them. It is as follows:

	RED.	WHITE.
Potassa,	20·80	30·17
Soda,	15·01	......
Lime,	1·83	2·76
Magnesia,	9·12	12·08
Peroxide of Iron,	1·29	0·28
Phosphoric Acid,	46·91	43·89
Silica,	0·15	......
Charcoal and Sand,	4·89	9·03
	100·00	98·21

All these constituents being derived directly from the soil, plainly indicate the reasons why our lands in the South are so easily exhausted. The crops extensively cultivated here all require, in a great measure, the same food from the soil; and hence soils which will not produce cotton, are alike incapable of producing the cereal crops. The great benefit derived from the application of cotton seed as a manure to these crops, is accounted for from the same causes; an abundance of phosphates being given in their application to the soil.

FALLOWING.—A system of tillage which carries away annually so large a proportion of these natural essentials to vegetation, and which provides no means of returning them, must necessarily impoverish any soil. A fixed principle in the agriculture of all countries where the prosperity of the future has at all been regarded, has been the gradual but certain improvement of the soil. This is necessary for the support of increased population; and in the Slave States, where there has been such an extraordinary and rapid increase of the laboring population, it should never be lost sight of. The intensity of our southern sunshine prevents, in a great measure, the annual coat of grass which supplies vegetable matter to the soil in northern climates; and the never-ending occupation of the soils, by our system of culture, prevents the natural improvement which in other countries is carried out by fallowing. We are well aware that fallowing is generally objected to in the South; and we think where fallow is converted into pasture land, and taxed during the whole season for the production of herbage to sustain greedy herds, the system might well come into disrepute. Planters, too, object to fallowing, and say they have not land enough to allow one-half to lie idle, &c.; but reason, and justice to the noble occupation of agriculture, allows this objection to pass unheeded; and its fallacy is proven by the desert wastes of "*old fields*,"—an agricultural feature only common to the New World, and, we blush to say it, only visible in the Southern or planting States. In Europe, where arable soil, compared to population, is a thousand times scarcer than in the Southern States, the agriculturists find fallowing a remunerative system. It is but little understood in American agriculture, and we may be pardoned for giving the proper details for fallowing, believing it to be the *cheapest* manner of renovating our soils. A field intended for fallow, should be deeply ploughed in mid-winter; the deeper the ploughing the better. This is simple preparation, but nevertheless necessary; and, above all things, keep every description of stock off in the field. The porousness of the soil will facilitate the assimilation of the natural salts of the earth, and atmospheric action, with the dissolving influence of the rains, will generally bring to the aid of the succeeding crop a sufficient quantity of these for its production. Late in autumn the herbage should be turned under. This process exerts chemical and natural influences beneficial to the soil,—First: as by decomposition of vegetable matter carbonic acid is produced, which is known to act as a powerful solvent of phosphated alkalies,—Secondly: those portions of the grass and weeds not readily decomposable, when admixed with the soil, give it that friability so necessary to easy tillage, and thus aids the agriculturist in his future labors. A *bastard* system of fallowing might, by the aid of the black and red tory pea, be judiciously adopted in the cotton-growing States. Owing to their imperviousness to wet, they can be sown in mid-winter, and, vegetating in the Spring without the aid of cultivation, generally make, upon ordinarily productive land, a sufficient crop to protect it from the sun in Summer, and smother out those weeds which are such a pest to cultivated crops. The constituents of the Indian pea—known to be in a great measure derived from the atmosphere—would in all probability furnish a better green crop for subversion, than the natural grasses and weeds. Judicious fallowing is, therefore, in our opinion, the cheapest, and by far the easiest mode of renovating and preserving the productiveness of our soils, and, if adopted and regularly persevered in, would heighten both the production and value of our cotton lands.

COMPOST MANURE.—Much may be effected in reclaiming worn-out cotton lands, by a good system of Compost Manuring; the benefits of which have been forced upon our Agriculturists by the gradual accumulation of animal manures, and the decomposition of wasted vegetable matter, in and around their barn-yards. It is a system which should be so generally understood and practised, that we deem it unnecessary to make other than a few remarks respecting the increase of this manure and its application. It is a mistaken idea, that the planter gains by hauling into his barn-yard, the stalks from his corn and cotton fields, in order to convert them into compost manure. Their elements would be returned to the soil, by the certain law of vegetable decomposition, if suffered to remain on the fields, and their place in the compost heap can be supplied easily by litter and leaves from the forests, grasses, weeds, and muck from the marshes, ditches, and fence rows on the farm. Weeds, abounding in the alkalies, furnish profitable vegetable matter for composting. In addition to these, we have the rotten wood and forest leaves, which are so abundant on all hands. Muck or peat, being decayed vegetable matter in mass, in this concentrated form contains a large amount of phosphates and alkalies—and when mingled with the droppings of animals, forms a compost highly retentive of substances thus imparted, which it yields most readily to the growing crops to which it is applied. Compost when applied in winter, does not require to be thoroughly decomposed, but when, as is the case on crops where it is applied in the spring, and its elements are demanded immediately by the young plants, its decomposition should be perfect. The compost heap should be protected from the rains, in order to prevent those salts rendered soluble by moisture, from being washed away. It would add much to the value of compost manure, if the water collecting on the roofs of farm buildings was carried in gutters entirely beyond the yard, and not allowed to flow through it, which would be greatly facilitated by a concentration of farm buildings.

Every domestic animal if properly confined and quartered, when not in use or grazing, would amply repay for the trouble in attending to it, and the filth from the wash house, stercorary, pig-pen, hen-house, and pigeon-cote, so much neglected amongst us, would if properly hoarded, furnish most valuable ingredients to the heap. A concentration of all that is essential to the production of our cultivated plants, being found in the component parts of this fertilizer—derivable from the cereal food consumed by animals, and the phosphate and alkaline properties of the weeds, grasses, &c., makes it at once the best and cheapest form of applying vegetable and animal manures for the immediate production of a crop, at the command of our planters. The quantity might be increased on every plantation in the State, to a degree which would make its manufacture profitable. This, however, will never be done until fewer acres are planted, which will enable them to manure more land.

Bone Manure.—Bones, according to Berzilius, contain 55 per cent., of the phosphates of lime and magnesia. The relative value of the bones of different animals varies in their constituents, and also from the difference in age, their value being increased with years. The bones upon every farm would furnish, if preserved and applied, a considerable amount of the best and most durable fertilizer, which is peculiarly adapted to the production of the cotton crop. This is proven by the identity of the constituents which compose bones, and are found in the cotton plant. The planter in the marl regions, especially where fossil bones and shells abound, has an abundant supply of native phosphate of lime, which only requires pulverization, to render it almost as useful as the recent bones. Phosphates in the bones comprise their chief value, which is shown by the fact, that they make a fertilizer equally as valuable, after the fatty matter has been extracted by soap boilers, as before—hence, all old bones might be rendered valuable if properly applied. Guano, the most powerful fertilizer applicable to husbandry, being the ordure of sea-birds, it is known, derives its great value from the amount of bone earth it contains. We therefore regard the annual waste of bones on plantations in the South, where more animal food is consumed than by any other people in the world, as the most suicidal disregard of that economy, which has furnished the axiom to agriculturists—"*that manure is wealth.*"

Many arguments abound to favor the adoption of bones as manure amongst us. One is, they can easily be preserved, and it only requires the same labor to do this that it does to throw them away. Another argument in their favor is, that a laborer, in a sack, can transport to a distant field, bone manure which will furnish more constituents to the crop, than can be concentrated in a four-horse load of the best stable dung, or compost manure—still another, is the little labor it requires to apply them to the soil. The great secret of applying bones to the soil, is found in pulverizing them into as finely separated particles as possible, which fits them for the operation of speedy atmospheric influence—in order that their constituents may be taken up rapidly by the plants. Grinding, crushing and burning, are the usual modes, but in order to fit the crushed bones or bone ashes for the greatest production, Professor Von Liebig, recommends the following process :

Pour over the crushed bones or bone ashes, half their weight of sulphuric acid, diluted with four parts of water, and after they have been digested for twenty-four hours, add one hundred parts of water—sprinkle this mixture over the field immediately before ploughing. By its action, in a few seconds the free acids unite with the bases contained in the earth, a neutral salt is formed, in a very fine state of division. Experiments instituted on soils, for the purpose of ascertaining the action of manure prepared in this manner, have distinctly shown that neither grain, nor kitchen garden plants, suffer injurious effects in consequence, but that, on the contrary, they thrive with much more vigor after its application. (*Vide Von Liebig's Organic Chemistry, American Edition, p.* 230.)

Another theory of application, by the great French chemist, M. Dumas, the substance of which we give from his article (contained in *Comptes Rendus, Nov.* 30, 1846, *p.* 1018,) "On the Manner in which Phosphate of Lime enters Organized Beings," is interesting. He remarks, that the phosphate of lime being insoluble in water, nevertheless penetrates, and is deposited in their structure, and bones containing it are slowly disintegrated by the soil and disappear after a time, under the influence of the rains. The investigations of M. Dumas discovered two causes producing these effects—the one acting rarely and feebly—the other constantly, and with great intensity.

The first resides in a property possessed by salammoniac, which facilitates the solution of phosphate of lime. Though this salt dissolves a notable quantity, and exists in all running waters—yet, this slight proportion renders its action in this respect inconsiderable.

9*

The second is found in the action of carbonic acid; and in this, the true solvent of phosphate of lime is to be found—for water impregnated with carbonic acid dissolves large quantities of phosphate of lime. M. M. Berzilius and Thenard, had remarked the alkalies and ebullition, by driving off, or neutralizing the carbonic acid which precipitated it.

M. Dumas, believing the action of the carbonic acid to be such as above stated, did not doubt the effect it would produce on the bones themselves. He therefore introduced plates of ivory into bottles of Seltzer water, (which contains a great deal of carbonic acid,) and they were as much softened in twenty-four hours, as if acted on by dilute hydro-chloric acid, which is also a powerful solvent of phosphate of lime. The Seltzer water was found loaded with phosphate of lime, and the experiment proved the action of carbonic acid as its solvent, to be both rapid and certain. I am sure this discovery will be of importance to the Agricultural world.

I would call the attention of physiologists to this property in carbonic acid, as it satisfactorily explains the transformation of the phosphate of lime into plants. Of course, it would not be practicable to dissolve the phosphate of lime by the aid of Seltzer water, but the preparation of bone ashes by its known and powerful constituent, might be rendered available in the following manner. Where bone powder or ashes is intended for manuring soil destitute of vegetable matter, let them be mixed with leaves or other organic matter, and its decomposition with the aid of the rains and atmospherical influence, will create a sufficient quantity of carbonic acid to assimilate the phosphates in such a form, that they will be readily taken up by the organism of the plants.

How easily could a planter manure a few acres of cotton with bone powder or ashes! When all the bones are hoarded as gold, and their true value known, they will be appreciated.

Then a bone mill for crushing, and simple apparatus for their chemical reduction, will be as essential to producing the crop, as a grinding mill is, to prepare grain for the food of man.

WOOD ASHES, containing phosphates and alkalies, to a considerable extent, where they abound, may be used advantageously as a manure for cotton.

LIME, being useful in decomposing and ameliorating adhesive soils, might be profitably employed in the permanent improvement of cotton lands.

Common potter's clay, diffused through water and added to milk of Lime, thickens immediately upon mixing, and if the mixture be kept for some months, and an acid be added, the clay becomes gelatinous, which is the effect of the admixture of the lime. The lime in combining with the elements of the clay liquifies it, and what is more remarkable liberates the greater portion of its alkalies. These interesting facts, so important to the scientific world, were first observed by M. Fuchs, at Munich, and led to the explanation of the effects of caustic lime upon the soil, which furnishes the agriculturist with an invaluable means of opening it, and setting free its alkalies—substances so indispensable to the production of his crops. (For further facts concerning lime, and its application to Agriculture, see Liebig's Organic Chemistry, which should be in the hands of every one.) The lime lands of the West producing abundant crops of cotton, so long as furnished with vegetable matter, shows that lime alone, upon exhausted soils, would prove a doubtful aid.

We could add suggestion after suggestion, relative to the aids to be applied to the production of cotton, upon exhausted soils, but these being the most important, we shall dispense with the boundless materials which lie abundantly around us, and only need transporting to our fields in order to benefit them. It was a matter of surprise to Professor Von Liebig, that any soil not furnished by artificial means with the preponderating constituents of the cotton plant and cotton seed, should produce a crop abounding in the phosphates. This leads me to further investigations, and a rich field of research still lies unexplored, in the analytical examination of the cotton soils of the South and West.

SECTION III.—REPORT ON THE ANALYSIS OF COTTON AND ITS SOIL.

OFFICE OF STATE CHEMIST, 29 EXCHANGE BUILDINGS,
BALTIMORE, Nov. 7, 1854.

THE following report on an examination and analysis of Sea Island Cotton Fibre and Seed, and the soil on which it grew, (the samples being carefully taken by the State Chemist on a late visit to Edisto Island,) was made for an intimate friend, (who owns large plantations of sea island,) in order to recommend a manure best adapted to the growth of this national staple; but inasmuch as a subject of such vast consequence as the increased production of the cotton plant should be placed before the country at large, I with pleasure accede to your request, and furnish you for publication with the analysis of the cotton fibre, cotton seed, and cotton soil, in order that a manure may be compounded adapted to the wants of the plant and corresponding to the deficiencies of the soil upon which it is cultivated, and that the benefits of its use may be extended as far as this variety of cotton is cultivated.

Any substance added to a soil to increase its products, without a knowledge of the constituents of the substance, the deficiencies of the soil, and the requirements of the crop, must depend for its success on mere accident and lucky guesswork; whilst a manure compounded with reference to the wants of the soil and nature of the crop grown on it, must be successful, because used on rational principles, and as a cause to produce an effect, having a direct connection with and dependent on it.

The cotton plant, like every other plant, requires for its perfection certain climate influences, proper cultivation, and a soil of proper physical texture, containing substances which do not and cannot exist in the atmosphere. All plants derive one part of their nourishment from the air, and another part, their mineral constituents, or ash, from the soil. Lime, magnesia, potash and soda, with various combinations of chloride, phosphoric and sulphuric acid, are necessary—*absolutely* necessary to the growth of the cotton plant. Without these no cotton plant has ever existed, and they cannot be obtained from the atmosphere, (with the exception of chloride and soda under particular circumstances,) and therefore they must either exist in the soil, or be supplied by the application of manure, or this plant will not grow. Manures, therefore, are nothing but substances necessary to the growth of a plant, which are deficient in the soil. If any soil contained all the substances which a plant required, in proper form for its use, there could be no manure for this soil, because there would be no deficiency to supply, and the plant grown on it would reach a degree of perfection limited only by the influence of its cultivation, and the climate. If on a soil containing all of these substances no manure would act, then on a soil deficient in any one of them, a manure would act only by supplying that deficiency, and should contain nothing but the substance deficient. All others would be useless.

To manure any soil, then, as a matter of course, its deficiencies should be ascertained, and the manure made with reference to those deficiencies. These deficiencies can be ascertained in two ways; the one by a long-continued course of practical experiments; the other by chemical analysis of the soil and plant. The world has depended, from its earliest ages, upon the first mode. With what success, the condition of agriculture, until the past ten years, will best answer. Improvement only reached a certain point, and that a very low one, and then ceased. Practical farming was as good in the days of Augustus as in the days of Washington. Farming was as well conducted in Italy eighteen hundred years ago, as in the United States or England, twenty years ago. Mere practice, then, without the aid of science, failed to lay down a rational manuring system for the growth of wheat or any other crop; and if practical experiment failed to do this for wheat in eighteen centuries, how much less has it done for cotton in fifty years? How much more will it do for it in the next five hundred? Agriculture, with the aid of both practical experiment and chemical science, has advanced, in the manuring of the wheat crop, more within the last twenty years than in the five hundred preceding it. Have we not, then, just reason for the belief that if so much has been done for the art of agriculture by the application of a science yet in its youth, its manhood will give results which we now do not dream of? Should not the cotton plant avail itself of this new aid to its culture and productions, and use the means which it affords with a liberal hand? * * *

The effect which we desire, is the production of the cotton plant in its greatest perfection.

The causes of production are the physical state of the soil, the climate, the cultivation of the crop, and, when required, manuring. We shall not speak of the physical character of the soil in this place; nor of the climate, because it is beyond our influence; nor of cultivation, because that can be best done by the owner by means within his own control.

What are the substances necessary to the growth of the cotton plant? Are all or any of these deficient in the soil? If so, then the manure best adapted to the soil is the one most abounding in the deficiencies of the soil, and such a manure must be recommended by the teaching of science, aided by all the lights of experience.

First. What are the substances necessary to the growth of the cotton plant which exist in the soil?

The cotton plant, like other plants, is composed of two grand classes of organs, one directly and the other indirectly tending to the perpetuation of the species; the first is the stalk and leaves, the second the seed and its appendage, the cotton fibre or wool.

The following table of the analysis of the Cotton Fibre and Seed, shows the composition of each, and the proportionate quantity of the substances which they require:

COTTON,

General Per Centage, Components of, as to

	FIBRE.	SEED.
Water,	4·72	9·51
Organic Matter,	94·03	86·46
Ash or Mineral Matter,	1·25	4·03
	100·00	100·00

Per Centage Composition of the above Ash or Mineral Matter.

	FIBRE.	SEED.
Potash,	35·26	34·75
Soda,	5·11	1·10
Lime,	16·73	6·00
Magnesia,	9·47	13·73
Peroxide of Iron,	2·07	0·55
Silicic Acid,	0·26	trace.
Phosphoric Acid,	5·42	35·85
Sulphuric Acid,	3·53	3·96
Chloride,	6·60	0.47
Carbonic Acid,	15·55	3·59
	100·00	100·00

The composition of these two will show what they require, and if their requirements be not allowed, they will fail to grow.

From the above it will be seen that the wants as to mineral matter of the cotton wool, or fibre, are chiefly potash and lime. Potash is the chief desideratum in a soil to produce the fibre. If the soil be deficient in this, then potash should be the chief constituent in the manure; this is a self-evident proposition. Next to this in quantity, we have lime; if the soil on which cotton is planted contains not this in sufficient quantities, then the manure should supply the deficiency. This is also a truism; because we know that neither potash nor lime is furnished to crops, except through the agency of the soil, or manures. Soda is also a component of the cotton fibre to a large extent; but we need not make this a constituent of a manure for this crop, because from the locality where it is grown, (near to the ocean shore,) a large quantity of soda, in the form of common salt, is supplied to all of the soils of these Sea Islands, in the spray from the ocean. Here then is a source of supply. The same is true of chlorine, which is here always associated with soda. Phosphoric and sulphuric acids likewise exist in the fibre. All of these are necessary to the full development of the cotton fibre; and without these it cannot exist. Not the least fibre could be produced unless on a soil containing not one, or several, but all of these constituents. So much for the cotton fibre as to its wanting of mineral constituents; furthermore, it requires a mechanical basis for its growth; there are seeds from which the fibre springs; without a healthy seed of strong vital power, the fibre will be small in quantity and of inferior quality. We now, therefore, turn our attention to it, and seek its wants from its analysis.

The analysis of the seed shows it to be much richer in

mineral matter than fibre; the latter containing only 1.25 per cent. of ash, whilst the former contains 4·03 per cent. In the seed the chief mineral constituent is phosphoric acid; more than one-third of all the mineral composition of the seed being composed of this; we have next in quantity potash, also composing more than one-third of the whole amount of mineral matter; next in quantity we have magnesia, then lime, then sulphuric acid; and as neither of these substances can be furnished by the air, if the soil be deficient in them, they must be supplied by manures. They are essential to the growth of the plant, and if not present in the soil in proper quantity, and suitable form for assimilation, the plant, without manure, will languish and die.

We thus are told by the fibre and seed, in plainest language, what they need for their full development; the cotton plant seeks this kind of food from the soil. Can the soil respond to its wants? Is it capable of furnishing all of the constituents shown in the above analysis in proper quantity, and in proper form, to supply what the plant needs? If the soil can do this, then no mineral manure is necessary. We will submit the soil to the same scrutiny as that to which the fibre and seed have been subjected. We will add to this, information derived from practical experience in manuring the soil—a thing never to be despised, and we will see in these two modes, each confirming and strengthening the testimony of the other, what should be the composition of the manure best adapted to the crop, and at the same time the wants or the deficiencies of the soil upon which it grows.

The soil upon which the above-examined cotton was raised, is composed, as to its bulk, of nine-tenths of fine alluvial sand, and of one-tenth of a cement consisting of sand, peroxide of iron, clay, lime, magnesia, and humus. It is not alone the proportionally very small quantity of cement to that of sand in which this soil differs from ordinary productive soils, known as clayey loams, and which renders it a very light one, of little tenacity, or of retaining powers for water, and nourishment in general, but this condition is also due to the nature of the cement itself, which does not show a proper quantitative proportion of its constituents. These constituents ought to be united to each other in such a proportion, that none of them can exercise a predominent influence. Sand, lime and magnesia, on one side, have to temper the tenacity and binding of the clay, iron, and humus, and by these means permit the free influence of the air upon the soil, and the rain water to penetrate it intimately without resting upon it. Clay, iron, and humus, on the other side, have to exercise their binding and water-reserving powers, but only to such an extent as will retain the solution of nourishing substances without doing injury to the porosity of the soil or its communication with atmospheric ingredients.

An examination of the cement of the soil in question, shows no such quantitative proportion of its constituents. It is almost entirely composed of sand, and peroxide of iron, next to those of clay, then of magnesia and humus, and only of such small quantities of lime as is quite common in soils. It is most probably to this fact that the larger per centage of magnesia must be attributed, which we meet in the composition of the ash of the cotton fibre, and especially in that of the seeds raised upon this soil. The want of lime induced the cotton plants to appropriate more abundantly magnesia, a substance which, in its chemical character and properties, stands nearest to lime, and which, therefore, is capable of substituting it to some extent. It is, however, beyond doubt that a substitution of lime by magnesia, induced by circumstances of necessity as they may have occurred here, will rather act injuriously on the quality of the fibre than improve it, and that, therefore, the planter's principal endeavor must be directed to the formation of a more calcareous cement, as well as it regards the mechanical texture of the soil as its directly nourishing properties. The improvement in the mechanical texture of this soil will best be effected by the application of a clayey marl, a substance composed of clay and lime, of which the former will increase the slight tenacity and water-reserving powers of the soil, whilst the latter will supply the present deficiency of lime.

If clayey marl cannot be procured, it may be best substituted by any kind of mud, the texture of which is stiffer than that of the soil, mixed with common oyster shell lime, which should be applied to this soil on the surface, and suffered to remain there as long as possible. This will act on the soil in the double capacity of improving its texture, and affording lime as a nutriment.

As to the directly nourishing properties of the soil, the analysis shows one acre, one foot deep, weighing 3,000 tons, to contain—of phosphoric acid, less than 15 lbs.; sulphuric acid, less than is contained in one bushel of plaster of Paris; chlorine, more than is contained in four bushels of common salt; potash, less than 20 lbs., a quantity so small that it could not accurately be ascertained; soda, more than four bushels of common salt contained.

We here, therefore, meet with—

A deficiency of phosphoric acid;

A deficiency of sulphuric acid;

A deficiency of potash;

and on the other side with—

An abundance of chlorine;

An abundance of soda.

As to soda, it stands nearest in its chemical character to potash, and though it is itself not nourishment for plants, to any great extent, as the quantity of it decreases in plants in proportion to their cultivation, it nevertheless acts as a substitute for potash, in the same manner as magnesia for lime. The composition of the cotton staple, as given above, shows the presence of soda in its ash in no small quantity. This circumstance seems to express, in accordance with the analysis of the soil, that by the scarcity of potash the plants were forced to assimilate soda. In this condition of things, the cotton plant could not be produced in its most perfect form.

If we now consider attentively—first, the requirements of the cotton seed and fibre; and secondly, the capacity of the soil to meet these requirements, we shall find the chief deficiencies to be these—first, a deficiency of lime in the soil; secondly, a deficiency of potash; thirdly, a deficiency of phosphoric and sulphuric acid.

This is taught us by the analysis. How is this borne out by practical experience in the manures of these soils? It is confirmed to the very letter. The best manure for the cotton plant is cow-pen manure and cotton seed—both rich in phosphoric acid and potash. The success of these manures proves a deficiency in the soil of their chief constituents, which are phosphoric acid and potash, otherwise they would not act as manures. The first of these sources, the cotton seed, cannot be obtained in sufficient quantities to manure all the land in cultivation, nor can cow-pen manure be applied to all of the soil in cultivation, because of the few cattle raised on the sea island cotton lands. We then must have recourse to a supply of manure, not directly the product of the soil itself.

This manure should especially contain phosphoric acid and potash, because they are the substances most needed by the cotton plant, and at the same time those least abundant in the soil. Practice has taught us, moreover, that in the great majority of soils, some nitrogenous manure is necessary, in order to give a quick early growth to the young plant. We must then apply a manure composed of nitrogenous compounds, phosphoric acid and potash; the constituents necessary can be very easily and cheaply supplied.

Peruvian guano is the cheapest source of supply of nitrogenous compounds.

Bone dust for phosphoric acid.

The various refuse of manufactories for potash.

Sulphuric acid is best supplied by plaster of Paris, which need not be used when bones are dissolved in sulphuric acid, and used as a constituent of the manure.

Whatever may be the productive capacity of cotton soil in its virgin state, it must deteriorate by long continued cultivation; this must be met by having the composition of cotton. The following table shows the substances and the quantity used in a crop respectively by the fibre and seed.

COTTON,

Production Per Acre.

Fibre,	200 pounds.
Seeds,	600 "

General Composition (in pounds) of

	200 lbs. Fibre.	600 lbs. Seeds.
Water,	9·44	57·06
Organic Matter,	188·06	518·76
Ash or Mineral Matter,	2·50	24·18
	200·00	600·00

Composition of the above Ash, as taken away by a crop from one acre (in pounds).

	BY THE FIBRE.	BY THE SEEDS.	IN ALL.
Potash,	0·881	8·403	9·284
Soda,	0·128	0·266	0·394
Lime,	0·418	1·451	1·869
Magnesia,	0·237	3·320	3·557
Peroxide of Iron,	0·051	0·132	0·183
Silicic Acid,	0·007	trace	0·007
Phosphoric Acid,	0·136	8·669	8·805
Sulphuric Acid,	0·088	0·958	1·046
Chloride,	0·166	0·113	0·279
Carbonic Acid,	0·388	0·868	1·256
	2·500	24·180	26·680

This table shows, that in order to maintain a soil in its original excellence, manures must be added having the composition of the cotton cultivated; and they must, for practical utility, not only contain all the constituents of the cotton, but have an excess to provide against loss from all sources which tend to the depreciation of manures.

The manure compounded by you for the cotton plant, is mainly composed of these ingredients, and must of consequence be peculiarly adapted to its growth, and the permanent improvement of the land upon which it is grown, since more of it is applied than is consumed by the plant.

I have been very cautious, for various reasons, to recommend no artificial manure unless guaranteed as to its composition, as the objection to many artificial manures is that they are not made of uniform character. This objection is met, in that which you sell, by the guarantee which you give; a thing done according to suggestions given in the Second Annual

Report of the State Chemist, to the House of Delegates, several years since. It is not necessary for us here to endorse either your pecuniary responsibility or your personal character. The former can be easily proven by any one who may become interested; the latter has been endorsed in various ways, at different times, by your party, by the people, and by our National Executive. First, by your party, in nominating you for the highly responsible and lucrative office of High Sheriff of Baltimore City and County—a nomination confirmed by a large majority of your fellow-citizens; then, by your appointment as a delegate to several of our National Presidential Conventions; and, subsequently, by your appointment to the second gift of the President, in our State—that of Naval Officer of the Port of Baltimore. In each of these instances you had for competitors some of the best men in our State. The best future recommendation for your manure will be its results, and to them we can look forward with implicit confidence.

There doubtless is a marked difference in many of the cotton lands of the South, which will have to be made known by a chemical analysis. Some may be deficient in one substance, and some in another. This variation, of course, is to be met by a corresponding change in the constituents of the manure; and its failure must be attributed to the peculiar nature of the soil, not the manure, when constituted as above recommended.

Wishing you success in your enterprise, and those who patronize it most abundant crops, we are very truly yours,

JAMES HIGGINS.
CHARLES BICKELL.

JOHN KETTLEWELL, Esq., Baltimore.

CHAPTER VI.

COTTON CONSUMPTION AND COTTON TRADE—COTTON TRADE FROM 1825 TO 1850. BY PROFESSOR McKAY, LATE OF THE UNIVERSITY OF GEORGIA.

SECTION I.—COTTON BAGGING.

WILL not the planters of Georgia encourage the use of bagging made from cotton? Listen to these facts and decide for yourselves.

The cotton crop of 1839, by the published statistics, was from Georgia, 163,000,000 lbs. Averaging the bag at 400 lbs., this made the crop 407,500 bags. This required, at five yards per bag, 2,036,500 yards, which at twenty cents per yard, is $407,500. If the bagging made from cotton be used in place of *hemp*, every dollar of this money is retained in the State; whereas, with the use of hemp, every dollar is carried out of it, except the small items of transportations and commissions. For safety's sake, we may say that $300,000 of this amount is taken away from the State entirely.

Again, to manufacture this bagging each yard requires two lbs. of raw cotton, which makes an amount of 4,077,000 lbs. Now if we use *hemp bagging*, we add just this amount annually to the *supply* from the crop for manufacturing purpose, and it tends to diminish the demand just so much. Suppose we convert it into bagging, we furnish a new *demand* for that amount. In other words, we withdraw from market that amount, diminish that portion of the supply—reduce the crop so much, which at 400 lbs. per bag is 10.192 bags, and thereby *increase* so much the *demand* for our cotton. In addition to the large amount of cotton thus consumed, there is also a considerable quantity converted into rope and twine. Will not the farmers study these facts, and take the hint? Read this article again, and see how you like my suggestions. May not we reduce the price of bagging to sixteen or seventeen cents, if we encourage entirely our own manufactures in making it, and save commissions, profits and freights now made by commission merchants, and ships and steamboat owners.

PUTNAM. [*Southern Recorder.*]

SECTION II.—COTTON BEDS—A GOOD SUGGESTION.

We find the following in the Albany *Cultivator.* Cotton beds are becoming very much in use on steamboats on the Western rivers, and they are considered superior to any kind but hair:

COTTON BEDS.—We have received from J. A. Guernsey, Esq., a copy of the *Southron,* published at Jackson, Miss., containing some remarks on the advantages of cotton for bedding. These advantages may be summed up as follows. It is claimed that " It is the cheapest, most comfortable, and most healthy material for bedding, that is known in the civilized world. In addition to these, may be named *superior cleanliness;* vermin will not abide it: there is *no grease* in it, as in hair or wool; it does not get *stale* and acquire an *unpleasant odor,* as feathers do; moths do not infest it, as they do wool; it does not pack and become hard, as moss does; nor does it become dry, brittle and dusty, as do straw or husk; and in many cases *medicinal.*" It is said not to cause that lassitude and inertia which is produced by sleeping on feathers. People not acquainted with it, have supposed they have been sleeping on the best feathers, when in fact their beds were made of cotton. The relative cost of cotton compared with feathers, hair, &c., may be seen from the following statement:

Cost of a Hair Mattress.—They are generally sold by the pound, and cost from fifty to seventy-five cents per pound. Thirty or forty pounds will cost $15 or $20.

Wool.—Thirty pounds of wool at thirty cents per pound, $9; twelve yards of ticking, at twelve and a half cents per yard, $1 50; labor, thread, &c., $2 75. Total, $13 25.

Feathers.—Forty pounds feathers, at thirty cents per pound, $12; fifteen yards of ticking, at twelve and a half cents per yard, $1 87½; labor, &c., $2 75. Total, $16 62½.

Cotton.—Thirty pounds cotton, at eight cents per pound, $3 40; twelve yards ticking, at twelve and a half cents per yard, $1 50; labor, thread, &c., $2 75. Total, $7 65.

It is recommended to run the cotton through a "picker," where one can conveniently be obtained, before using. This gives it additional cleanliness and buoyancy.

The substitution of cotton for bedding throughout the United States would be an immense saving, beside opening a new avenue for that article, to an extent, according to the estimation of this writer, equal "to more than two of the largest crops of cotton ever produced in the United States."

SECTION III.—A NEW USE FOR COTTON.

A late number of the New York *Day-Book* contains the following notice of a novel application of our great southern staple:

"Invention, which goes far to make useful almost every production of nature, has found a new use for cotton, in which, without doubt, a very large amount will be employed. We allude to the mattresses now coming so favorably and extensively into use, in preference to any article heretofore tried. The writer of this has used one for six months past, and has fonnd it to possess every requisite and desirable quality of a mattress, without the objections so frequently urged against moss, curled hair or husks—as the husks moulding from damps, bad smells from the curled hair in summer, and the lumpy matting of the moss. The cotton felting, prepared by a patented process, has none of these annoyances, is always elastic, and will, with ordinary care, last a lifetime. Our friends, 'way down on the old plantations,' will please make a note of this, and consider that the invention is a feather in their caps—or rather money in their purses—as the demand for the raw material at home will, doubtless, materially increase the price. We feel sure that if the real qualities of this mattress are ever made known to the public generally, five hundred thousand bales a year would not satisfy the demand for its manufacture. The article having been thoroughly tried on the principal steamships, and approved by their owners, as well as by physicians who have tried, and strongly recommend them, we doubt not the patentee will make a fortune on them. The agents for this city, and the Union generally, are Messrs. Doremus & Nixon, 21 Park place and 19 Murray street."

SECTION IV.—DOMESTIC BAGGING AND BLANKETS.

Mr. Jones:—Being a reader of the *Southern Cultivator*, I have noticed several articles on the subject of the manufacture of cotton bagging, from cotton; and having some practical knowledge on the subject, I avail myself of the invitation given in your paper, to give the result to the public through its columns, with the hope of contributing something to render the planters more independent, and add to their comfort.

During the late war with England, when bagging was scarce and very high—one dollar per yard—I purchased a sley, made for the express purpose of weaving bagging, forty-two inches wide, sixteen "*bier.*" At that time I grew flax, and made several pieces of bagging, as good as the imported article, which I sold for one dollar a yard. Twelve or thirteen years ago, I put up a screw for packing cotton ; since which I have made all my bagging, rope, and twine of cotton, which was spun upon a common wheel, and wove in a common loom in the old sley. The warp should be sufficiently stout to work two threads in a reed, and the filling coarse enough to keep the cloth the full width of the sley. Every girl that can draw a thread, can spin the filling; and if the warp and filling are sufficiently stout, the bagging will be of as good quality as may be desired. I make mine to weigh from $1\frac{1}{2}$ to $1\frac{3}{4}$ pounds per yard, and in four and a-half yards I pack 330 to 350 pounds of cotton. I always save the inferior cotton to make the bagging and rope, and my bales look as neat as any bales I see in the market, and generally, I believe, command as good a price as any, according to the quality of the staple. None need fear of success if the thread be sufficiently stout, and the sley be of the right kind. I therefore think that almost all planters, particularly small planters, may supply themselves with an excellent article of bagging and twine of their own make.

There is another article I find greatly to my advantage to manufacture at home, which is wove in the same sley and loom,—I mean blankets for servants. If made heavy, and well wove, they will be as warm as, and much more durable than the imported article. They may be made of any convenient size; and if the filling be slack twisted and neatly carded after being wove, the blanket will do no discredit to the bed of a gentleman and his lady. My family have made their blankets for many years, and those wives and daughters who feel a desire to excel in the manufacture of their carpeting, will find such a sley a great auxiliary: and so far as the home manufacture of these articles can contribute to our independence and comfort, we can accomplish it; and I hope that our wives and daughters have both industry and patriotism sufficient to make the effort, provided they are seconded in their exertions by their husbands and fathers.

Yours, respectfully,

Gwinnett Co., October, 1843. ANSELM ANTHONY.

SECTION V.—COTTON RIGGING FOR SHIPS.

THIS article, we are glad to see, continues to grow in public favor. The *Delta* states that at one time, during the month of April, there were the following ships—all new, and of large tonnage—in the port of New Orleans, with a part or the whole of their running rigging and hawsers of cotton cordage:—North America, Escort, Shakespeare, of Boston; Knickerbocker, of New York; Erie, Liberty, St. Patrick, of Thomaston, Me.; Walter Scott, Civilian, Saniscott, Robert Lane, Sea Breeze, Sewell, of Boston. The officers of all these ships were unanimous in their testimony in favor of cotton cordage for running rigging, and many of them thought it would be adopted for standing also.

The *Delta* states that Donald McKay, the celebrated ship-builder at Boston, the owner and builder of the famous clipper ship Republic, is adopting cotton cordage for all his new ships. The large new clipper ship Caleb Cushing, recently built at

Newburyport, Mass., has all her rigging, both standing and running, of cotton cordage. Capt. J. P. Smith, of the ship Walter Scott, gives it as his opinion that it will outlast any rope, whether hemp or Manilla. He is also quite sure the cotton rope is the strongest of the three ropes, as by bending cotton and Manilla ropes of equal sizes together, and heaving on it, at the capstan, the Manilla will always part first. Capt. Brown, of the ship Escort, says that he has used cotton cordage twenty-eight months on the ship Medora, and found it to wear far better, on all accounts, than any other rigging he ever used. In wet weather, likewise, it is more pliable, and in frosty weather it is not so stiff as Manilla. After it is used a few months it becomes smooth and glossy, and works through the blocks much better than any other rope. After the Escort was launched last autumn, at Bristol, Me., she was made fast with two Manilla lines, and three and a-half inch line of cotton cordage seventy fathoms in length, and a very heavy blow came up and the two Manilla lines parted, and the ship rode for more than twenty-four hours, and during the gale, with this line, run out its whole length, alone to hold her; and the strain was so great that it wore and imbedded its full size into the white oak crosstrees, without breaking a thread in it. It is Captain Brown's opinion, that no Manilla or hemp rope of the same size could have held the ship under like circumstances. A number of shipmasters' statements, all to the same purport as the above, are published in the *Delta*, all going to show that cotton cordages, like cotton duck, is destined to come into general use.

SECTION VI.—PAPER FROM THE BARK OF COTTON.

We called attention some months ago to specimens of hemp made from the bark stripped from cotton stalks and left at our office for public inspection. We now learn from the *New York Day Book*, that specimens of bark have been exhibited to paper manufacturers at the North, which is found to be of a fibrous character, and is considered to be well adapted for the manufacture of good paper.

The best period for preparing this cotton hemp will be as soon as practicable after the picking of cotton has been finished. The plants should then be pulled up and dew-rotted like hemp or flax, and afterwards broken up and the bark separated from the wood of the stalk. The specimens of clean bark exhibited to experienced paper makers was considered equal to good rags worth six cents per pound, or about $120 per ton, and was pronounced the best substitute for rags of any raw vegetable material known to the trade.

The importance of an abundant and cheap material as a substitute for rags, from which good, cheap paper can be made, may be judged of from the fact that the United States consume as much as France and England combined. There is no element in the progress of civilization more important than cheap paper.

For some years, the consumption of paper has been gaining on the supply of rags, and fears have been felt that the advance in their cost would ultimately be seriously felt in every department of literature, so that should the discovery of cotton hemp realize the anticipations of paper makers, it will not only prove valuable to the South, but also to the civilized world.

The magnitude of the paper business may be conceived when we take into consideration that there are seven hundred and fifty paper mills in the United States, employing three thousand engines, and which produce annually, at ten cents per pound, $27,000,000 worth of paper. To manufacture this amount of paper requires 405,000,000 pounds of rags,

1¼ lbs. of rags being necessary to produce one pound of paper. The value of the rags, at the average of four cents per pound, amounts to $16,000,000, to which, if the cost of making them into paper, including 1¾ cents to each pound of paper in labor, with wastage, chemicals, &c., be added, would swell the cost to $23,625,000 to produce $27,000,000 of paper—leaving nett profits on the total manufacture of $3,375,000. For the year ending the 30th June, 1855, we imported 40,013,516 pounds of foreign rags from twenty-six different countries. Of this amount, Tuscany, in Italy, supplied 14,000,000 pounds, Two Sicilies 6,000,000, Austria 4,000,000, Egypt 2,466,928, Turkey 2,466,928, England 2,591,178. The total value of the 40,-013,516 pounds imported was $1,225,150.

It is possible that the cotton fields of the south may furnish an almost inexhaustible supply of hemp, so as hereafter we will reach the great desideratum in modern civilization, an abundant and cheap supply of paper.—*Savannah Republican.*

SECTION VII.—COTTON-SEED OIL.

The Wakulla *Times* of a late date, says:—"'The proprietors of one of our linseed oil mills have commenced the manufacture of oil from cotton seed, and about 400 bags of the seed arrived here this week from Memphis, to be used for this purpose. The oil is used for burning. How far the parties will succeed in their enterprise remains to be demonstrated. We believe the manufacture of oil from cotton seed has been carried on in the South to a greater or less extent for several years; at Natchez, we believe, one of these mills has been in operation for some ten years, but, so far, the oil has not come into general use. The difficulty seems to be in clarifying, as it will not burn in a crude state. Should our enterprising citizens succeed in preparing the oil for use, it will prove a most important article of commerce.'—*Cincinnati Price Current.*

"Perhaps there is now more cotton seed oil used for table and other purposes than even consumers themselves are aware of, to say nothing of the soap, which is of a superior quality, made from the refuse of the oil after clarifying. On this subject, a friend, whose statements may be relied on, writes us:

"'I notice in a Western paper that a concern in Cincinnati has commenced the manufacture of oil from cotton seed. I will mention a few facts which may be of use to somebody. There is a prejudice against cotton seed oil, but it is owing mainly to the fact that the seeds have been extensively used for that purpose without hulling—the hull imparting to the oil a bitter taste and a gummy substance, which injured it for drying, and causes a smoke when burning. Notwithstanding this, quantities of this oil have been mixed with linseed and lard oils, and the buyers have been none the wiser for it. Some three years since, a friend of mine commenced the manufacture of oil from cotton seed. The seeds were first perfectly hulled so that nothing but the meat of the seed was used.

"'After the oil was extracted, it went through a clarifying process (a simple one, but very perfect), leaving it as clear and as pure as the best olive. For burning it has no superior, as it gives a clear, brilliant light, without smoke; and for the table it can scarcely be surpassed, for it has deceived and is still deceiving many good judges of the article. Indeed, my friend assured me that he was unable to fill all the orders for oil put up for the table—but he added: "We dare not call it cotton seed oil lest it might prejudice the sale."'

"We of the cotton-growing States can safely feel ourselves perfectly independent of the world for oil for all purposes."

SECTION VIII.—COTTON SEED AS A MANURE.

MR. EDITOR:—The great enriching properties of cotton seed as a manure, and its superior power of imparting an early impetus to the growth of plants, have been visible to all who have ever given them a fair trial. They need not be confined as a manure to any one article grown by the planter, but extended to almost every species of vegetation—corn, peas, cotton, vegetables, small grain, and grapes—though not equally beneficial to all alike. From an experience of a few years, the subscriber would advise their use on land designed for corn, in the quantity of seventy-five bushels to the acre, to be hauled out after the land is well fallowed, a few days before planting time, and deposited in piles of equal quantity and distance for convenience and facility in distributing them. The land being suitably prepared and ready for planting, the rows laid off by a shovel-plough, opening broad and deep, the seeds are then scattered from one end of the row to the other, with the corn dropped on them at such distance in the drill as the quality of the land will justify, say in medium or average land two and a-half feet apart, covered with a turning-plough, and harrowed off with an iron-tooth harrow. If the corn be planted and seed sown on it, the stand will be greatly endangered from the lint and heating quality of the seed, but by planting as advised, a stand will be secured. If a greater quantity of seed can be procured, the benefit desired will be more general and permanent to the land by scattering them broad-cast, and ploughing them in. Many contend that this manure is not felt longer than one year, but such persons, after exposing the seeds all the winter, haul them in small piles and suffer them to remain from March until May, when they are removed to the corn hill, there deposited on the surface around the stalk to remain uncovered until wind, rain and sun dissipate its fertilizing properties. My own impression is, that its influence is felt for five years, independent of an increased quantity of vegetable matter returned to enrich the land. If the season proves suitable, by this plan of manuring in the drill, one may realize a hundred per cent. in the yield of corn; and the succeeding year, if planted in cotton, in reversing the beds, this very manure is thrown on the bed where the seeds are sown, enabling the plant to reap early benefit at a period, as generally acquiesced in by planters, when it requires more support than at any other, in order that its early growth and healthy condition may enable it to escape the ravages of lice, with which the plant is never attacked until enfeebled by cold or some other injurious cause. We are urged by many to manure exclusively for cotton. From such I think differently. The past year, ten acres were in cotton, where a hundred bushels of cotton seed were given to the acre, placed in the water furrow, and bedded upon them. The result was an increased growth, and moderate increase in yield, but not enough to justify such an expenditure of this valuable manure. The same year fifty acres were planted in cotton that had been grown three preceding years in corn and peas, manured each year with cotton seed, as advised, but none on it the year it was in cotton. The corn stalks had been cut up, and with the pea vines, regularly turned under. The land in both cuts was well cultivated, and seasons alike. The soil of the ten acres was good, of a mulatto color, whilst that of the fifty acres was poor and hilly, with clay near the surface. The difference in the yield was fifty per cent. in favor of the fifty acre cut. This year I have ten acres planted in cotton with a hundred bushels of seed sown broad-cast and ploughed in; also, fifty acres, planted in cotton, which was in corn and peas the past year, manured with a hundred bushels to the acre in the drill, but none this year; the corn stalks cut up in several pieces, and with the pea vines turned under. Both cuts of land are similar in quality, and have been cultivated alike with like seasons. The result so far shows that the benefit derived from last year's manuring is greatly preferable to that of the present year. The stand on the ten acres is very imperfect and very irregular in its size, and has been much harrassed by lice. The fifty acre cut is a good stand, quite regular in its size, has been free from lice, and presents altogether a thrifty appearance, and bids fair to yield fifty per cent. per acre more than the ten acre cut.

From these remarks, you will readily perceive that I prefer manuring with cotton seed for corn, instead of cotton; that we are better rewarded the second year to succeed it in cotton, and well compensated the first for our trouble. I do not pretend to say that the benefit is altogether attributable to the cotton seed, but to the change in the crop, together with the advantage of corn stalks and pea vines, restoring the original susceptibility in the land to grow and produce good cotton. If those who disagree with me will give results from a better process of using this valuable manure, I shall be greatly obliged.

Prairie Mount, Miss., 1848. COWLES M. VAIDEN.

SECTION IX.—FEEDING HOGS WITH COTTON SEED.

FOR five or six years in succession, I fed my hogs with raw cotton seed. My plan was this: I put out such a quantity that each hog would have the measure that a shelled ear of corn would fill, of the seed, and gave at the same time an ear of corn to each hog. While the larger hogs were eating this, the pigs fed more fully on corn in a pen that the large hogs could not enter. I am not aware that my hogs, in any instance, ever sustained any injury from this course. They ate them freely, and from some comparative experiments, I think they kept in better condition than others that had the same allowance of corn, without the cotton seed. I say, I think, for the experiment was not very carefully made. I have also fed them boiled, and again I think without injury. My conclusion is, that with corn, hogs may safely have a small (equal) allowance of cotton seed. At the same time I am fully convinced they are very injurious to pigs; but managed as above, I never noticed any injury. I noticed that the hogs macerated, and sucked the pulp from the seed, and dropped the hull and lint upon the ground—perhaps pigs do not do this. Would not hulling free them from any injurious quality? Cannot some one answer?

SECTION X.—COTTON SEED.

Mr. Editor: Will you, or some of your correspondents, please inform me what would be the cost of a mill for extracting the oil from cotton seed? It is not very certain that it would be advisable for a cotton planter to manufacture oil from his seed, even if he could make it a profitable business, for they constitute one of the most valuable of our manures; but I should like to know what the profit would be. It would, at any rate, add to the interest of your columns if you could furnish your readers with an article on this subject, stating the cost, and *modus operandi* of manufacturing the oil.

I should like also to know if any of your subscribers have ever made the experiment of feeding hogs upon cotton seed, and what was the result. I made the experiment once, or rather my hogs did it without my knowledge or consent, of feeding them on raw, unrotted seed; they died in consequence; but I have no doubt a fine article of food for hogs might be prepared from them.

TETRARCH.

We have no information upon which we can rely, as to the cost of a mill for making oil from the cotton seed, or of the profit of the operation. It is a question of interest in a cotton-growing country, and one which some of our patrons are doubtless prepared to answer. We hope they will do so. We have often heard of the value of cotton seed when heated or partially rotted as food for hogs. We never had much faith in the recommendation, and therefore never tried it. It was perhaps because we knew that they were good for corn, and that corn was, beyond all question, good for hogs. There may, however, be more in it than we have imagined, and if so, there can be no harm in finding it out. Who can tell?—[Ed.

SECTION XI.—FEEDING SHEEP ON COTTON SEED.

Messrs. Editors:—Experience and observation has prepared me to believe that sheep which are fed on cotton seed are more subject to the rot and other diseases than when fed on other food. For the last eight years my sheep were wintered entirely on cotton seed; during the most of that time they were affected with a most distressing cough and running at the nose, which foretold their condition; and after they were turned to grass in the spring, running at large, they continued to cough and run at the nose, and when the weather became warm, would sicken and die in large numbers. This season I have fed entirely on fodder and oat straw, which they eat kindly, and in keeping them in this way I find they are now healthy and sound, free from cough and as clean about the nose as a goat.

Now, Messrs. Editors, if cotton seed feed produces the above stated facts, cannot some of your numerous correspondents, or Dr. Lee, enlighten the readers of the *Cultivator* on the subject.

I am, with respect, yours, &c.

Raytown, Ga., Feb. 1855. AARON W. GRIER.

SECTION XII.—COTTON SEED AS FOOD FOR STOCK.

ALTHOUGH cows, sheep, and hogs are very fond of cotton seed as a food, still I regard them as a bad and very unsafe article of provender. They will certainly kill hogs, grown ones as well as small ones, when eaten in an unrotted, or uncooked state. Some persons contend, however, that if well rotted, or cooked, they make an excellent article of food. From my experience and observation, I am satisfied that they are not good food under any circumstances. They are worse for hogs than for any other stock. Hence I never give them to my swine. As a substitute for hay, fodder, shucks or straw, I frequently give them to my cows and sheep through the winter. But I would never use them to feed any stock whatever, unless on account of a scarcity of the foregoing articles.

J. A. T.

SECTION XIII.—THE COTTON TRADE, FROM 1825 TO 1850.

BY PROFESSOR M'KAY, UNIVERSITY OF GEORGIA.

INSTEAD of our annual review of the cotton trade, for a single year, we propose to extend our examinations back to a longer period. For this purpose we have collected, in our statistical tables, the production, consumption, stocks, and prices of cotton, for each year from 1840 to 1850; and, for the more important particulars of the trade, we have gone back as far as 1825. This period of twenty-five years we have divided into intervals of five years, and given the average for each, noting the rate of increase or decrease for each country separately. By taking average results, we get clear of the fluctuations arising from short crops, and other disturbing causes, and are able to observe the general progress, free from those temporary variations which prevent our judging accurately the real changes that are taking place. In this review, we shall see a very prominent place assigned to our country. The United States is now, not only the largest producer, but the largest consumer, of cotton: our production has advanced with such rapid strides, that we have distanced all competitors: the cotton goods worn by our people exceed now the amount used by Great Britain and all her dependencies in the four quarters of the globe; and the demands of our manufactories have increased with much greater rapidity than those of any other country in the world. In the table of supplies, (Table I., at the end of this article,) we may observe, that, while other countries have been nearly stationary, our production has advanced with great rapidity. In twenty years, our average crop has increased from 848,000 bales, to 2,351,000, or nearly three hundred per cent. If the period of twenty-five years, from 1825 to 1850, be divided into five equal intervals, the increase for each will be found to be 27, 37, 38, and 15 per cent. In the same time, the production of all other countries has only risen from 383,000 to 440,000 bales; having absolutely declined, in the last five years, over 16 per cent. In the first period of five years, the crop of the United States constituted 68 per cent. of the whole; in the second, 74; in the third, 77; in the fourth, 80; and, in the fifth, 84 per cent. of the whole. As our bags have increased very much in weight, and are now much larger than those of other countries, our advance has been still greater, and our rank still higher than these figures indicate. If the table of consumption (Table II.) be examined, it will appear that our progress is none the less rapid, in comparison with other countries. In the same twenty years, the deliveries to our manufactories have advanced 325 per cent., viz., from 127,000 bales to 539,000; while, in the same time, the advance of Great Britain has been only 125 per cent., viz., from 653,000 bales to 1,472,000. In each one of these periods, our rate of progress has been more than twice as rapid as hers; and though the absolute amount of our consumption is yet far below that consumed by the English manufacturers, yet, in the last five years, our increase has been 176,000 bales, while theirs has been only 180,000. At present, our consumption is 37 per cent. of the English, while twenty years ago it was only 19 per cent.

France, during all this period, has remained nearly stationary. Twenty years ago, her consumption was 257,000 bales; now, it is only 363,000. In the last five years, she has gone backward, the decline having amounted to 58,000 bales. From 1825 to 1830, the deliveries to her manufactories were double those of the United States; now, they are 33 per cent. less than ours. Her rank, compared with Great Britain, and with nearly every other country in Europe, has also declined. In Spain, Belgium, Germany, Holland, and Russia, the increase has been nearly as rapid as in the United States. In the last five years, their advance (Table III.) has been 46 per cent.; ours, 49 per cent. Their rank in the cotton-consuming countries is yet low, but their rapid progress will soon bring them to a more important position. At present, their consumption is 34 per cent. of that of Great Britain, and the time is not far distant when, taken together, will equal her. Twenty years ago, the comparative rank of the United States, Great Britain, France, and the rest of the continent, was in proportion to the numbers 11, 55, 22, and 12; in the last five years, the percentage of each has been 19, 51, 13, and 17. If France be left out of the comparison, the rank of each, twenty years ago, was as 13, 70, and 17; now, it is as 21, 59, and 20. Although Great Britain requires for her manufactories more than half of all the cotton worked up in Europe and America, the amount actually used by her people, including all that is exported to India, British America, Australia, and all the colonial dependencies of Great Britain, is less than the amount used in the United States. This has been shown to be true for the last four years; and the present year, although it exhibits an apparent decline in our home consumption, forms no exception to this result. The enlarged imports of cotton goods imported into our seaports, compensate, in part, for the falling off of the wants of our factories. If we compare the progress in the demand and supply, it will be seen that, during the last five years, the consumption has increased much faster than the production—the one having advanced 19 per cent., and the other only 9. This might be inferred from the decline in the stocks; but it will be more satisfactory to consider the average production and consumption of the last ten years. The average amount taken by the manufacturers, from 1840 to 1845, was 2,414,000 bales, and, from 1845 to 1850, 2,869,000 bales, showing an increase of 465,000 bales; while the supply advanced from 2,561,000 bales to 2,791,000, with an increase of only 230,000 bales. When it is remembered that the last period embraces the year 1848, when, from the revolutions in Europe, the consumption declined over 600,000 bales, and the years 1845 and 1849, when the American crop so far exceeded its usual average, this result will be more striking and important. The table of stocks (Table IV.) confirms and establishes this same result. At the end of 1844, the cotton on hand in Europe was 1,101,000 bales; at the end of 1849, it was only 646,000 bales.

It may be further observed, that the increase in the supply during the last five years, has been slower than the natural increase of laborers. The advance in the one has been only 9 per cent., and in the other 12 or 13. As many new hands have been brought to the Southern States during this period, the rate of increase in the working force of the cotton-growing States has been still greater than 12 or 13 per cent. This excess has occurred at no former period. From 1825 to 1850, the increments for each period of five years have been 18, 32, 33, and 9 per cent.; always above the increments of population, except in the last interval. It follows, from this, that labor and capital have found other modes of employment more attractive and profitable than the raising of cotton. It is well known that this has been to some extent true in the United States, but it has been more evident and striking in India and Brazil. In these countries, the crop has declined 16 per cent. in the last five years. From Brazil, it has declined regularly for the last twenty years; and the recent advance in coffee will tend still more to divert labor from the production of cotton. The abolition of the discriminating duty in favor of East India cotton, by Sir Robert Peel, and the very low prices which have recently prevailed, have not only stopped any increase in the imports of Surat and Madras, but turned the current in the opposite direction. The advance in the fifteen years before 1845, was 10, 80, and 60 per cent., in each interval of five years; but, from 1845 to 1850, the decline has been 24 per cent. It may fairly be deduced from this, that the prices of the last five years have not afforded sufficient encouragement to production, and that the planters may now look for a permanent improvement in prices. The table of prices (Table V.) shows that for the last five years the average price at the seaports of the United States has been seven cents and three mills; and it may be expected, with confidence, that they will not rule so low hereafter—that the average rates will not merely experience a temporary rise, as if caused by the short crop and the small stocks of the present year, but a permanent and continued advance.

The table of stocks (Table IV.) represents the amounts on hand in the seaports of Europe continually increasing from 1840 to 1850, while during the four years ending 1849, they have been nearly stationary. Comparing them with the wants of the manufacturers, as is done in the column which contains the number of weeks that the stocks would supply the consumption of the factories, the supply was a trifle lower at the close of 1849, after the receipt of the largest crop ever brought to market, than it had been during the last ten years. The number of bales was a little greater than at the close of 1848; but the time this stock would supply the wants of the manufacturers, was a little less. After this review of the history of the trade in cotton for the last ten years, if we remember that the production of 1850 has been much below the average of the last five years, and that the prospects of the next year's crop are but a little better, it is evident that the present advance in cotton is founded upon no speculative basis, but on the unchangeable laws of supply and demand. Two short crops are succeeding each other, while the stocks on hand are very much reduced. To this it may be added, that everything is favorable to a large consumption. Peace everywhere prevails, except in the unimportant Duchies of Schleswig Holstein. Money is abundant, and the currency everywhere undisturbed. Food is very cheap. The present harvest of Europe, as well as the last, is much above an average. Thus, while stocks are low, and the supply small, the demand is large. Prices, therefore, must maintain a high level, unless commotions in France, or some unforeseen event of commanding importance, interferes with the regular operation of commerce. In considering the supply and demand of the coming year, we must, therefore, base all our estimates on high prices. The receipts from India and Brazil, and the consumption in Europe and America, will all be affected by this fact. If the advance were slight, it would not experience any sensible check; but when the price has risen to its present rate, (13½ cents for middling fair, Savannah, October 23d,) an advance of 85 per cent. over the average of the last five years, the amount purchased even in our country may be expected to decline. The supply for 1851 will probably exceed that of 1850, not only from the United States, but from India and Brazil.

The past season here has been unfavorable for the growth of cotton; but its disasters, especially in the West, have not been as severe as in the preceding year. In South Carolina and Georgia, there will be a decided decline. The late cold spring, and the long drought in June and July, left the plants small, and the bolls few and scattering. The severe storm on the 24th of August blew out on the ground much open cotton, and prostrated and twisted the stalks so much, that there has been no late crop of forms to mature in October. September was a beautiful season for gathering, and so was much of October. There are some plantations where the crop is very fine. The hot summer favored a rapid growth, and repaired, in part, the injury done by a late spring. The general drought was, at some places, relieved by local showers, which brought out some superior crops. The amount of land planted was greater than ever. The receipts at Charleston and Savannah will also be increased by the extension of the Georgia Railroad to the Tennessee river. Were it not for this last cause, a falling off of 100,000 bales might be anticipated. With this, the deficiency will not probably exceed 70,000 or 80,000; and the receipts of these two ports may be expected to reach 650,000 bales. From Alabama, the reports have not been so disastrous. The spring was late, and the stand poor; but the dry summer prevented the ravages of the worm, which had done so much damage the preceding year. The river floods had also done harm the last season; and these they have escaped. The prairie-lands have not suffered so much with rust as before. On the Tombigbee, and also on the Black Warrior, the prospects of the planters are very much above those of last year. On the Alabama, the promise is about the same as last year. Still the disasters have been severe, and the crop will be below an average. An increase of 90,000 or 100,000 bales in the receipts at Mobile, including the Montgomery shipments to New Orleans may, with confidence, be anticipated. From Florida, a slight increase may be looked for. The amount of land planted has been considerably enlarged, and the drought has not been as general as in the eastern part of the cotton region. At New Orleans and in Texas, a gain may be looked for. The failure last year was so great, that it is almost impossible to expect a like deficiency again. From Louisiana, Arkansas, and the greater part of Mississippi, the reports have been better than last year. The early frost of October 6th, injured not a little of the cotton as far north as Memphis; but in general, even in Tennessee, the plant remained green and flourishing, till the general frost at the close of the month. The production of Tennessee and North Alabama will fall below that of last year, and a portion of this will not reach New Orleans. The crop was everywhere backward, but the hot, dry summer helped to repair this damage, and by keeping off the caterpillar and boll-worm, permitted the forms to mature. The severe storms that did so much harm in Florida and the Atlantic States, did not extend so far to the west. The season for gathering has been very fine, and the time of frost late enough to mature nearly every boll that could make cotton. The average receipts at

New Orleans, for four years past, has been 943,000 bales; and this period includes two short and two full crops. For the present year, I would estimate them at 850,000 bales. Combining these estimates, the whole supply from the United States will amount to 2,200,000 bales (see Table VI.), which is about 100,000 in advance of the last five years. The receipts from India have increased very much during the present year, under the stimulus of high prices, and they are destined to advance still more for the coming season.

The purchases now making in Bombay for the English market are reported to be large; and when the new crop begins to arrive at the seaports, the current will turn still more strongly towards England. Not only is their production enlarged by high prices in Europe, but a larger portion of the crop is diverted from China, and from domestic use, for the Western markets.

The average imports into Great Britain for the last three years have been 211,000 bales; but for the first nine months of the present year, they have reached 128,000 bales for Liverpool alone; and for the whole year, for all the ports, they will probably reach 300,000 bales. For 1851 not less than 325,000 bales may be anticipated. This is higher, much higher than any former year. The year 1841 was the largest before 1850, and then the amount was 275,000 bales. The high prices that are now prevailing, and that are likely to prevail for the present season, authorize us to expect an increase, even over the present year. (Table VII.)

From Brazil, Egypt, and other places, an advance over the usual average may also be looked for. The average imports into England from 1845 to 1849, were 175,000 bales; but for the present year, the amount will exceed 260,000 bales, and for 1851 will be still larger. (Table VIII.) If we estimate them at 275,000, the whole supply from all these sources

(Table IX.) will reach 2,800,000 bales. In reference to the consumption, we may remark, that the purchases for our home manufactories have declined during the present year over 30,000 bales. The high price of the raw material, the low duties on foreign goods, and the immense imports of cotton fabrics from England, have caused this retrograde movement. In 1849, there was a falling off of 14,000 bales, so that our consumption is now 44,000 bales below that of 1848. Doubtless the stocks in the hands of the manufacturers are very small, and a slight advance in goods would set all the mills at work again. The universal prosperity of the country forbids us to expect the extension, or even the continuance of this depression. For 1851, I would estimate the demand at 500,000 bales, which is 11,000 above the consumption of the present year (Table X.), and 13,000 below the average of the last year. In Great Britain the falling off in the purchases of the manufactures have been very slight (Table XI.), and as the reported purchases last year were 80,000 or 90,000 bales above the actual deliveries to the manufacturers, the real deficiency is less than the apparent. For the present year, the consumption in Great Britain will not be below 1,500,000 bales, against 1,588,000 in 1846, and 1,491,000 in 1848. Everything has been favorable to a large consumption, except the price of the raw material. Money has been abundant—food of all kinds cheap—and labor well rewarded. These elements of prosperity have not been confined to Great Britain, and therefore her exports of cotton-goods have been unprecedentedly large. The home and foreign demand being both good, the factories have run full time, in spite of the high price of cotton. This never occurred before, and cannot be expected again, with any considerable confidence. At every former period, an advance in the raw material has checked the demands of the factories, and lessened the purchases of the consumers. For the coming year, everything is fully as favorable as the last; and if these favorable tendencies have counteracted the tendency of high prices in the raw material, it will be proper to expect the same for 1851 as for 1850. We may, therefore, set down 1,500,000 bales as the probable English consumption for the next year.

In France there has been a decided decline (Table XII.) in the deliveries to the manufacturers. Our exports have fallen from 368,000 bales to 290,000, and the stocks on hand the 1st of October, were almost exactly the same as last year. The purchases at Havre for the first nine months of the present year have been 249,000 bales, against 290,000 in 1849. From these figures we cannot estimate the consumption of American cotton for the present year higher than 300,000 bales, against 351,000 for 1849. No advance on this can be expected for the next year, nor is there any reason to anticipate any appreciable decline. For the rest of Europe, we have the exports from the United States for the present year, 194,000 bales, and the exports from Liverpool up to October 11th, 193,000 bales. The whole English exports of 1849 were 254,000 bales; and as their amount on October 12th was 21,000 more this year than last, the whole exports for the year from all the ports will probably reach 275,000 bales, making the total supply from these two countries of 469,000 bales. As the stocks on hand on the continent last year were very low, it is impossible to reduce them much lower. They are now, however, at several ports, lower than last year, so that the consumption will probably exceed 469,000 bales. As this is a decline of over 100,000 bales from 1849, it is not to be expected that so low a limit can be reached for the year 1851. Heretofore their progress has been forward and rapid; and were it not for high prices, this would continue. If we estimate their wants for 1851 at 500,000 bales, we have the total consumption (Table XIII.) of 2,800,000 bales—the same as the supply. As the stocks are now much lower than last year (Table XIV.) and as they were then very low, they will bear no further reduction without a material advance in prices. On the contrary, any decline in price would immediately permit the consumption to expand, not only in France and the rest of the continent, but even in England. We may expect, therefore, that the present high range of prices will be maintained.

The review that has been taken of the supply and the demand, shows that the present advance in cotton is the result of no speculative movement, but that it is based on the immutable laws of trade. The long prevalence of low prices has stimulated consumption and diminished production, until the stocks on hand have fallen to an extremely low limit. Exactly at this point an unfavorable season has lessened the crop, and an abundant harvest, and every other element of general prosperity, have encouraged the demand. We congratulate the planters on the handsome returns they are receiving for their crops, and we may extend our congratulations to the whole country, for what benefits them is a benefit to all.

TABLE I.—*Supply of Cotton, (in thousand bales.)*

YEARS.	U. S. Crop brought to Seaports.	U. S. consumed in the South.	Total United States Crop.	East India Imports into Great Britain.	Brazil, etc., Imports into Great Britain.	Brazil, etc., Imports into other places.	Total besides United States.	Total, all kinds.
1840	2178	50	2228	216	146	111	474	2701
1841	1635	55	1690	275	166	128	569	2259
1842	1684	55	1739	255	124	166	545	2284
1843	2379	60	2439	182	165	176	523	2962
1844	2030	60	2090	134	197	80	511	2601
1845	2395	65	2460	150	201	105	461	2921
1846	2101	70	2171	95	155	69	319	2490
1847	1779	80	1859	254	135	122	481	2340
1848	2348	90	2438	228	137	36	401	2839
1849	2729	100	2829	182	245	111	538	3367
Average from 1825 to 1830	838	10	848	73	211	99	382	1231
" " 1830 to 1835	1055	20	1075	81	186	108	375	1450
" " 1835 to 1840	1440	35	1475	144	196	104	444	1919
" " 1840 to 1845	1981	56	2037	232	160	132	524	2561
" " 1845 to 1850	2270	81	2351	177	175	88	440	2791
Increase per cent. in 20 years	171	....	177	142	17	....	15	117
" " 15 years	115	....	119	118	8	....	17	99
" " 10 years	58	....	59	23	11	....	1	40
" " 5 years	15	....	15	24	9	....	16	2

TABLE II.—*Consumption of United States, Great Britain, France, and of Europe and America, (in thousand bales.)*

YEARS.	United States, n'rth of Richmond.	United States, Total for.	Total for Great Britain.	United States Cotton in France.	Total for France.	Total for these three.	Total for Europe and America.
1840	295	345	1271	374	440	2056	2370
1841	297	352	1158	368	422	1932	2252
1842	268	323	1207	364	442	1972	2310
1843	325	385	1385	351	409	2179	2573
1844	347	407	1438	335	392	2237	2564
1845	389	454	1574	351	419	2447	2918
1846	423	493	1574	360	403	2470	2968
1847	428	508	1131	252	293	1932	2296
1848	532	622	1491	276	303	2416	2901
1849	518	618	1588	351	399	2605	3264
Average from 1825 to 1830	117	127	653	....	257	1037	1187
" " 1830 to 1835	175	195	876	....	269	1340	1540
" " 1835 to 1840	240	275	1069	....	349	1693	1943
" " 1840 to 1845	307	363	1292	....	421	2076	2414
" " 1845 to 1850	458	539	1472	....	363	2374	2869
Increase per cent. in 20 years	290	325	125	....	41	129	142
" " 15 years	161	176	68	....	35	77	86
" " 10 years	91	96	38	....	4	40	48
" " 5 years	59	49	14	....	14	14	19

TABLE III.—*Consumption of Europe and America, omitting England, France, and United States, (in thousand bales.)*

YEARS.	Exports from United States.	Exports from Great Britain.	Direct Imports from Egypt.	Stock, Jan. 1.	Stock, Dec. 31.	Consumption.
1840	182	123	49	72	112	314
1841	106	116	74	112	88	320
1842	132	138	88	88	108	338
1843	194	119	118	108	145	394
1844	144	141	23	145	26	287
1845	85	122	37	126	99	471
1846	205	194	26	99	26	498
1847	169	215	81	26	87	404
1848	255	192	9	87	58	485
1849	322	254	63	58	88	659
Average from 1840 to 1845	....	....	....	....	....	338
" " 1845 to 1850	....	....	....	....	....	495
Increase per cent. in five years	....	..	...	..	...	46

TABLE IV.—*Stock 31st December, (in thousand bales.)*

YEARS	Liverpool.	Great Britain.	Week's Consumption in Great Britain.	Havre.	France.	Rest of the Continent.	Whole of Europe.	Week's Consumption.
1840	366	464	18	80	97	112	673	17
1841	430	538	24	90	135	88	761	21
1842	457	561	24	109	138	108	807	21
1843	654	786	9	101	125	145	1056	25
1844	745	897	32	53	78	126	1101	26
1845	885	1057	35	52	65	99	1221	26
1846	439	547	18	25	47	26	620	18
1847	364	451	16	43	53	87	591	17
1848	393	498	17	20	31	58	587	18
1849	468	559	18	38	49	88	646	18

TABLE V.—*Amount, Value, and Price of American Cotton.*

YEARS.	Exports in millions of pounds.	Value, in millions of dollars.	Price of Exports.	Whole crop of United States.	Value of United States.	Liverpool prices of Uplands in pence.
1840	744	64	8·6	891	77	6
1841	530	54	10·2	684	70	6¼
1842	577	48	8·1	704	58	5⅝
1843	817	49	6·0	988	59	4⅝
1844	664	54	8·1	857	69	4⅞
1845	873	52	6·0	109	61	4⅜
1846	548	43	7·9	901	71	4⅞
1847	527	53	10·1	771	78	6¾
1848	814	62	7·6	1011	77	4¼
1849	1027	66	6·5	1174	76	5⅛
Average from 1825 to 1830	219	28	12·8	288	37	7¼
" " 1830 to 1835	312	34	10·9	387	42	7¼
" " 1835 to 1840	446	64	14·4	560	81	8¼
" " 1840 to 1845	666	54	8·1	825	67	5½
" " 1845 to 1850	754	55	7·3	972	71	5⅛

TABLE VI.—*United States Crop.*

	RECEIPTS.			ESTIMATE.
	1848.	1849.	1850.	1851.
Texas, bales	40,000	39,000	31,000	50,000
New Orleans	1,191,000	1,094,000	782,000	850,000
Mobile	436,000	509,000	351,000	440,000
Florida	154,000	200,000	181,000	190,000
Georgia	255,000	391,000	344,000	300,000
South Carolina	262,000	458,000	384,000	350,000
Other places	10,000	28,000	24,000	20,000
Total	2,348,000	2,729,000	2,097,000	2,200,000

TABLE VII.

English Imports from East Indies.

	Imports. Bales.	Remarks.
1835 to 1840, average,	144,000	High prices.
1840 to 1845, "	232,000	Chinese war.
1845 to 1850, "	177,000	Peace, low prices
1848, Oct. 6, to L'pool.	93,000	Moderate prices.
1849, Oct. 5, "	69,000	Low prices.
1850, Oct. 4, "	128,000	High prices.
1848, whole year,	223,000	Moderate prices.
1849, "	182,000	Low prices.
1850, est. whole year,	300,000	High prices.
1851, " "	325,000	High prices.

TABLE VIII.

English Imports from Brazil, Egypt, etc.

Years.	About 1st Oct. Liverpool. Bales.	Whole yr. Gt. Brit. Bales.
1846,	121,000	155,000
1847,	75,000	185,000
1848,	94.000	137,000
1849,	178,000	245,000
1850,	208,000	260,000
1851,		275,000

TABLE IX.—*Supply of C[illegible].*

	1849.	1850.	1851.
	Bales.	Bales.	Bales.
Crop of United States	2,729,000	2,097,000	2,200,000
English imports from East Indies	182,000	300,000	325,000
" " other places	245,000	270,000	275,000
Total from these sources	3,156,000	2,667,000	2,800,000

TABLE X.—*United States Consumption.*

Years.	Amount Consumed.	Average for 3 years.	Inc. per cent. per annum.	Inc. per cent. for 3 years.
1846	423,000	386,000	9·0	23
1847	428,000	413,000	7·0	32
1848	532,000	461,000	11·5	30
1849	518,000	493,000	7·0	28
1850	488,000	515,000	4·0	24

TABLE XI.—*Deliveries to the Trade at Liverpool.*

	1849.	Consumption each week.	1850.	Consumption each week.
	Bales.	Bales.	Bales.	Bales.
March 8	324,000	36,000	227,000	25,222
April 12	433,000	30,929	338,000	24,143
May 10	562,000	31,222	501,000	27,833
June 21	748,000	31,167	672,000	28,000
July 6	835,000	30,926	742,000	28,222
August 9	1,037,000	32,206	907,000	28,942
September 6	1,141,000	31,694	981,000	28,029
October 4	1,220,000	30,500	1,836,000	27,150
November 11	1,287,000	31,390	1,116,000	27,219

TABLE XII.—*Deliveries to the Trade at Havre.*

	1849.	Consumption each month.	1850.	Consumption each month.
	Bales.	Bales.	Bales.	Bales.
May 1	120,141	30,035	104,728	26,182
July 1	193,971	32,328	167,653	27,942
August 1	243,040	34,720	200,650	28,664
September 1	279,541	37,442	332,190	29,024
October 1	290,585	36,323	249,707	27,523

TABLE XIII.—*Consumption.*

	1849.	1850.	1851.
Great Brit., all kinds	1,588,000	1,500,000	1,500,000
France, of Am. cotton	351,000	300,000	300,000
The rest of the Cont't	596,000	470,000	500,000
Total	2,535,000	2,270,000	2,300,000

—*Hunt's Merchants' Magazine.*

TABLE XIV.

Stocks at Recent Dates.

	1849.	1850.
	Bales.	Bales.
Liverpool, Oct. 12	583,000	482,000
Havre, Oct. 9	46,000	46,000
United States, Sept 1,	155,000	168,000
Hamburg, Oct. 1	5,000	2,000
Total	788,000	698,000

SECTION XIV.—COTTON STALK HEMP.

From the Madison Family Visitor.

Our readers will be interested in the extract which follows, from a letter of Col. John B. Walker, to the Editor of this paper, under date New Orleans, Nov. 11th. The Cotton Stalk Hemp promises to be an article of quite considerable importance.

"I went yesterday with Gen. Gordon and Lieut. Governor Horton, of Texas, to see the Cotton Harvest Gatherer, and the Cotton Stalk Hemp. The first article is intended to pick cotton from the boll and put it in a bag. This I regard as worthless—not a humbug, because it will not likely deceive any one but the inventor: farmers will not be caught with it. The Cotton Stalk Hemp is, in my opinion, worthy of the highest consideration. It has the color of the Gunny or East India bagging, and the fibre is as strong as that of the hemp. It is prepared by knocking off the lateral limbs of the cotton stalk, then cutting down the stalk and burying it in a plough furrow in the field, where it lies covered up for fifteen days. It is then taken up and broken, as you break hemp, and this clears it of the woody fibre, and it is fit for spinning in any way, either by machinery or by hand. Again, it is better prepared by sowing the cotton broadcast and thickly; this causes the stalk to run up in height and clear of lateral limbs, more nearly resembling the hemp weed.

"The discoverer of this process of making cotton stalk hemp is a Frenchman, of Baton Rouge, by the name of John Blanc. He has been four years engaged in experiments, and has now just obtained his patent right from the Patent Office.

"I was happy to meet our old county friend, Mr. Wm. J. Vasoon, at the place where this cotton stalk hemp is exhibited, and see him test the strength of the fibre. When we shall get the cotton wool, and then the bark from the stalk for our bagging and rope, and the oil from the seed, and the cotton seed hulls converted to some practical purposes, and the cotton stalk *roots* manufactured into patent medicine as an elixir to perpetuate the existence of the negro who cultivates the plant, we can then imagine that it has its true, intrinsic and inestimable value. It will then be worth a war on the part of the South to sustain and defend it, and claim a place for it and its cultivators in any reputable portion of this earth."

CHAPTER VII.

LETTER FROM THE SECRETARY OF STATE, TRANSMITTING A STATEMENT RESPECTING THE TARIFF DUTIES AND CUSTOM-HOUSE REGULATIONS APPLICABLE TO AMERICAN COTTON.

DEPARTMENT OF STATE,
WASHINGTON, May 30, 1856.

SIR: In compliance with the resolution of the House of Representatives of the 12th instant, "that the Secretary of State be requested to communicate to this House, in tabular form, such information as may be in possession of the Department of State respecting the tariff duties and custom-house regulations applicable to American cotton in the principal commercial countries; also, tabular comparative statements showing, 1st. The quantities of cotton exported from the United States to the principal commercial countries, respectively, and the aggregate amount of duties derived therefrom; 2d. The quantities of cotton imported into Great Britain, France, and Spain, respectively, and the countries whence imported; 3d. The quantities exported by Great Britain to all countries, respectively; and 4th. The quantities and values of cotton manufactures and yarns exported from Great Britain and the United States; respectively, to all countries; each of these statements embracing a period of five years, from 1851 to 1855, both inclusive, or for so much of said period as authentic data are attainable ; together with such other general information respecting the cotton trade of the United States as may be deemed pertinent to the purport of this resolution," I have the honor to transmit the accompanying papers.

I have the honor to be, sir, your obedient servant,

W. L. MARCY.

Hon. N. P. Banks, Jr.,
Speaker of the House of Representatives.

Statistical Office, May 28, 1856.

Sir : I have the honor to submit to you, herewith, an answer to the resolution of the House of Representatives of the 12th instant, requesting certain information on the subject of the cotton trade, with the preparation of which I was charged. The subject has been treated in detail in the report on the commercial relations of the United States with all foreign nations, recently transmitted from this office, and now in course of printing.

I have the honor to be, sir, very respectfully, your obedient servant, EDMUND FLAGG, *Superintendent.*

Hon. W. L. Marcy,
Secretary of State.

I.—*Statement respecting the tariff duties and custom-house regulations applicable to American cotton in the principal commercial countries.*

COUNTRIES.	LBS.	RATES OF DUTY.
Great Britain, . .	——	Free.
France, ...	220,	In national vessels, $3 72; in foreign vessels, $6 48.*
Spain, ...	101,	In national vessels, 79½ cents; in foreign vessels, $1 25.
Russia, ...	36,	18¾ cents.
Bremen, ...	Ad val.,	⅔ of 1 per cent.
Sardinia, ...	——	Free.
Belgium, ...	——	Free.
Austria, ...	——	Free.
Sweden and Norway,	——	In Sweden, free; in Norway, nearly ½ cent per pound.
Mexico, ...	101,	$1 50.
Hamburg, . . .	Ad. val.,	½ of 1 per cent.
Holland, ...	——	Free.
Two Sicilies, . .	192·05,	$8.
British N. A. possess^{ns}.	——	Free.
Denmark, ...	——	Free.
Portugal, ...	101,	2 1-5 cents.
Tuscany, ...	——	Free.
Papal States, . .	74·86,	10 cents
Cuba, ...	101,	In national vessels, 19½; in foreign vessels, 27½ per ct. on a valuation of $5.

* By the treaty of 1822 United States vessels are equalized with French vessels in the direct importation into France of articles the growth, manufacture, or produce of the United States.

II.—*Tabular comparative statement showing the quantities of cotton exported from the United States to the principal commercial countries respectively, and the annual average amounts thereof; and the annual average amounts of duties derived therefrom, for a period of five years, from* 1851 *to* 1855, *both inclusive.**

Countries to which exported.	Pound of cotton exported from the United States in the years					Annual average amounts of cotton.	Annual average amount of duties paid.†
	1851.	1852.	1853.	1854.	1855.		
Great Britain	670,645,122	752,573,780	768,596,498	696,247,047	673,498,259	712,312,141	Free.
France	189,164,571	186,214,270	189,226,918	144,428,360	210,113,809	173,829,584	$2,939,300 25
Spain	84,272,625	29,301,928	86,851,042	35,024,074	83,071,795	83,704,292	265,296 06
Hanse Towns	16,716,571	22,188,228	22,671,782	37,719,922	30,809,991	26,011,298	‡25,795 00
Belgium	16,835,018	27,157,890	15,494,442	13,980,460	12,219,553	17,037,472	Free.
Austria	17,309,154	23,948,434	17,968,642	14,961,144	9,761,465	16,789,767	Free.
Sardinia and Italy	10,320,406	17,934,268	17,487,984	12,725,830	16,087,064	14,911,110	Different rates.
Russia	10,098,448	10,475,168	21,286,563	2,914,954	448,897	9,044,806	47,018 36
Mexico	845,960	6,700,091	7,463,851	12,146,080	7,527,079	6,936,612	§ 103,018 99
Holland	5,508,670	10,259,042	7,088,994	6,048,165	4,941,414	6,759,257	Free.
Sweden and Norway	5,160,974	5,939,025	6,099,517	9,212,710	8,428,437	6,968,132	‖Different rates.
British N. A. possessions	23,525	16,582	12,295	72,790	883,204	201,679	Free.
Denmark	....	37,042	435,169	32,983	209,186	142,876	Free.
Cuba	113,572	294,852	196,392	250,633	9,620	173,014	2,355 42
Portugal	....	98,235	87,691	121,059	144,006	90,198	19 64
Elsewhere	722,473	141,803	652,395	1,946,895	270,822	746,918	
Total to all countries	927,237,089	1,093,230,639	1,111,570,370	987,833,[illegible]	1,008,424,601	1,025,659,165	

* The data for this statement are derived from the United States treasury reports, in which the commercial year closes June 30. The year in British and French official documents corresponds with that of the calendar; hence one cause of apparent discrepancies in figures, for nominally the same years.

† The amounts of duties paid are calculated on the customs rates given in the preceding statement [I.], although those rates, during the five years designated, have, in some instances, undergone changes. Belgium, for example, did not admit cotton free until the passage of the law of April 12, 1854.

‡ The amount is calculated on the medium of the ad valorem duty of Bremen and Hamburg, on an assumed valuation of seventeen cents per pound.

§ The amount is calculated on the rates of the existing tariff of January 31, 1856, prior to which cotton was either prohibited or subjected to a duty equivalent to prohibition.

‖ United States treasury reports do not give quantities to Norway distinct from those to Sweden. In the latter, cotton is free; in the former, the duty is nearly half a cent per pound.

III.—*Tabular comparative statement showing the quantities of cotton imported into Great Britain, and the countries whence imported, for a period of five years, from* 1851 *to* 1855, *both inclusive.**

Years.	Pounds of cotton imported into Great Britain from						
	U. States.	Brazil.	Egypt.†	E. Indies.†	W. Ind.†	E'where.	All countries
1851	596,638,962	19,339,104	16,950,525	122,626,976	446,529	1,877,653	757,379,749
1852	765,630,544	26,506,144	48,058,640	84,922,432	708,696	3,960,992	929,782,448
1853	658,451,796	24,190,628	28,353,574	181,848,160	344,060	2,078,562	895,266,780
1854	722,151,360	19,708,600	23,353,120	119,829,152	205,072	2,090,800	887,333,104
1855	679,264,096	24,577,952	32,622,688	145,218,976	No data.	8,476,160	890,159,872
Aggregate	3,422,136,758	114,317,428	149,388,547	654,445,696	1,699,357	17,984,167	4,359,921,953
Average..	684,427,351	22,863,485	29,867,709	130,889,139	‡ 424,839	3,596,833	871,984,390

* Made up from British official authorities. The commercial year in England begins January 1; in the United States, July 1; hence seeming discrepancies in figures for apparently the same period of time.

† Egypt includes Turkey, Syria, and the Mediterranean generally; the East Indies include British India generally; the West India islands belonging to Great Britain, and British Guiana.

‡ Four years.

IV.—*Tabular comparative statement showing the quantities of cotton imported into France, and the countries whence imported, for a period of five years, from* 1851 *to* 1855, *both inclusive.**

Years.	Pounds of cotton imported into France from			
	United States.	Elsewhere.	All countries.	Value.
1851	127,418,053	19,083,961	146,402,014	$21,204,000
1852	171,235,021	†17,742,078	188,977,099	27,528,000
1853	178,608,904	19,537,722	198,146,626	28,830,000
1854	174,929,557	15,319,242	190,248,799	27,900,000
‡1855	....	....	....	....
Aggregate	652,191,535	71,683,003	723,774,538	105,462,000
Average	163,047,884	17,920,751	180,943,635	26,365,500

* Compiled from *Tableau Général du Commerce de la France.*

† Of which amount 11,973,427 pounds were from Egypt and Turkey, and 930,516 pounds from Brazil.

‡ No data.

V.—*Tabular comparative statement showing the quantities of cotton imported into Spain, and the countries whence imported, for a period of five years, from* 1851 *to* 1855, *both inclusive.*

The statistical office has no official Spanish data from which to make up the statement required.

The quantities of cotton exported from the United States to Spain, according to United States Treasury reports, the years specified, were as follows:

1851 . . .	34,272,625 lbs.	1854 . . .	35,024,074 lbs.
1852 . .	29,301,928 "	1855 . .	33,071,795 "
1853 . . .	36,851,042 "	Avrg. (5 yrs.)	33,704,292 "

From Cuba, the same years, according to "Balanzas Generales" of that Island, the quantities exported to Spain were as follows:

1851	13,145 lbs.	1854 . . .	1,489 lbs.
1852 . . .	300,225 "	1855	No data.
1853	138,625 "	Average (4 yrs.)	113,438 "

From Porto Rico, according to official Balanzas of that island, as follows:

1851	315,083 lbs.	1854 . . .	No data.
1852 . . .	141,807 "	1855 . . .	No data.
1853	245,552 "	Average (3 yrs.)	234,147 lbs.

From Brazil, according to the "*Proposta e Relatario*" of that empire for the years 1852–'3, and 1853–'4, the quantities of cotton exported to Spain were as follows:

1852–'53, ...	2,291,578
1853–'54, ...	2,351,279
Average, (2 years)	2,321,428

Spain, according to the "*Cuadro General*" of that kingdom for 1849, imported that year, from countries of production, 26,136,881 lbs. of cotton; of which quantity the United States supplied 21,669,441 lbs., Cuba 3,371,830 lbs., Brazil 832,-604 lbs., Porto Rico 370,881 lbs. and Venezuela 21,316 lbs.

VI.—*Statement showing the quantities of cotton exported by Great Britain to all countries, respectively, and the countries whence imported, for a period of five years, from* 1851 *to* 1855 *both inclusive.**

Years.	Exported to all countries.	Of which was imported from—				
		U. States.	Brazil.	Egypt.	East Indies.	Elsewhere.
	Pounds.	*Pounds.*	*Pounds.*	*Pounds.*	*Pounds.*	*Pounds.*
1851.......	111,980,400	66,921,344	1,888,880	211,008	42,959,168	......
1852.......	111,875,456	69,217,120	3,619,840	124,656	38,864,672	49,168
1853.......	148,569,680	82,701,472	4,786,768	948,416	60,082,064	50,960
1854.......	125,554,800	55,101,200	1,438,192	369,600	68,645,808	......
1855.......	124,345,760	56,989,632	759,360	386,064	66,210,704	......
Annual average.	124,465,219	66,186,153	2,493,603	407,948	55,352,483	......

* Compiled from the monthly "Accounts relating to Trade and Navigation," presented to the British Parliament, the only authority at hand from which the countries whence the cotton exported was imported could be ascertained. Results gathered from these monthly accounts sometimes vary from those given in the "Annual Statement of the Trade and Navigation of the United Kingdom," from which latter document was made up the table that follows.

Countries to which exported.	Pounds of cotton exported from Great Britain in the years—				Annual average.
	1851.	1852.	1853.	1854.	
Russia	35,185,472	45,605,840	48,937,392	208,544	32,484,312
Sweden	2,434,656	3,591,840	4,414,368	5,866,560	4,076,856
Prussia	1,576,064	674,240	1,143,296	23,444,624	6,709,556
Hanse Towns	27,473,040	22,472,016	33,417,440	36,055,264	29,854,449
Holland............	22,119,104	15,834,224	28,676,592	26,934,544	23,391,116
Belgium	12,856,480	12,657,680	18,466,672	14,040,768	14,505,400
France.............	1,365,504	2,225,440	2,403,968	2,759,232	2,188,536
Sardinia	2,742,320	2,238,208	3,860,864	3,821,328	3,165,680
Austria............	1,366,064	1,957,088	3,830,288	4,811,856	2,991,824
Other countries.....	2,647,120	2,324,560	3,418,800	5,383,392	3,443,468
Aggregate	109,765,824	109,581,136	148,569,680	123,326,112	122,810,688

NOTE.—No data for the year 1855.

VII.—*Tabular comparative statement showing the quantities and declared values of cotton manufactures and yarns exported from Great Britain and from the United States, respectively, to all countries for a period of five years, from* 1851 *to* 1855, *both inclusive.**

Years.	GREAT BRITAIN.				UNITED STATES.			
	Manufactures.		Yarns.		Manufactures.		Yarns.	
	Quantities.	Values.†	Quantities.	Values.	Quan's.	Values.	Quan's.	Values.
	Yards.	*Dollars.*	*Pounds.*	*Dollars.*	No	*Dollars.*	No	*Doll's*
1851..	1,543,161,789	110,246,010	143,966,106	33,246,010	data.	7,203,945	data.	37,260
1852..	1,524,256,914	108,242,290	145,478,302	33,273,275	..do..	7,637,433	..do..	34,718
1853..	1,594,592,659	119,509,700	147,539,302	34,478,265	..do..	8,746,300	..do..	22,594
1854..	1,692,977,476	116,884,300	147,128,498	33,456,985	..do..	5,486,201	..do..	49,315
1855..	1,935,846,937	130,623,375	165,499,547	36,152,140	..do..	5,857,181	..do..	None.
Agg'te	8,290,835,775	585,5[illegible]5,675	749,611,755	170,606,625	.....	34,931,060	.....	143,887
Aver	1.658.167.155	117.101.115	149.922,351	34.121.325	....	6,986,212	.....	28,777

* Made up from British and United States official documents, respectively; the commercial year of the former ending December 31, and of the latter June 30.

† The pound sterling is computed at five dollars.

VIII.—*General information respecting the cotton trade of the United States.*

GREAT BRITAIN.

The annual average importation of cotton from all countries into England, the last five years, has been 838,335,984 lbs. of which amounts, according to British authorities, 661,529,220 lbs., or more than three-fourths, were from the United States. The annual average exportation to the continent and elsewhere, has been 122,810,688 lbs., or about one-sixth of the total quantity imported, leaving 715,525,296 lbs. for the annual average consumption. About one-sixth of the whole amount imported was from British possessions.

In 1781 Great Britain commenced the reëxportation of cotton to the continent and elsewhere. In 1815 the quantity thus reëxported had risen from the annual average of 1,000,000 lbs. to that of 6,000,000 lbs. In 1853 the aggregate amount exported exceeded 148,500,000 lbs., of which nearly 83,000,000 lbs. were derived from the United States, and more than 60,000,000 lbs. from the East Indies. The quantity of American cotton reëxported by Great Britain to the different markets of Europe, when compared with the quantities imported, is much less than of that imported from other countries—a fact which suggests the superiority of the American article, and its better adaptation to purposes of fabrile industry. For example, about one-tenth of the cotton imported from the United States is reëxported, against nearly one-half of that imported from the East Indies. A comparison between American and East Indian cotton shows a difference of 100 per cent. in favor of the former; the cotton of the East Indies containing 25 per cent. of waste, while that of the United States contains only 12½ per cent. The fibre, also, of the latter excels that of the former.

In 1788 the efforts of the East India Company commenced for the promotion of the growth of cotton, and for the improvement of its quality, in British India; and the first exportation of the article to England was made the same year. In 1814 the exportation amounted to 4,000,000 lbs.; it now averages 165,000,000 lbs. per annum. An area of about 8,000 square miles is said to be devoted to the culture.

Liverpool is the great mart of the cotton trade of Great Britain, and of Europe generally. Thus, while the total imports of that article into the United Kingdom, according to British authorities, in 1852, amounted to 2,337,338 bales, the quantity at this port reached 2,205,738 bales. About six-sevenths of the cotton received at Liverpool comes from the

United States; and of this four-fifths is estimated to be imported for the factories of Lancashire and Yorkshire.

Since March, 1845, cotton has been admitted into British ports free of duty. Prior to that period the duty was, of and from British possessions, 8 cents; from other places, 70 cents per 112 lbs.

The number of spindles in operation in England is estimated at more than twenty millions.

The value of cotton supplied by the United States to Great Britain in 1855 was $57,616,749, being about the average each year the last four.

The quantity of cotton exported from the United States to England, in eleven months of the fiscal year 1856, is estimated at 2,755,000 bales.

It appears from "Commerce and Navigation," that the importation of raw cotton from the British West Indies into the United States has increased, for some years past, in a ratio quite proportional to the decrease of such importation into Great Britain. Thus, the importations of cotton into the United States and Great Britain, respectively, from the British West Indies, from 1851 to 1855, inclusive, were as follows:

1851	Into the U. S.	29,353 lbs.	Gt. Britain.	446,529 lbs.
1852	"	6,756 "	"	703,696 "
1853	"	252,892 "	"	344,060 "
1854	"	159,381 "	"	205,072 "
1855	"	880,217 "	"	No data.

The average price per pound of cotton, from 1851 to 1855, inclusive, in the United States and Great Britain, respectively, is shown as follows:

Average Price per Pound.

1851	In the U. S.	12·11 cents.	In Gt. Britain.*	$12\frac{1}{4}$ cents.
1852	"	8·05 "	"	$11\frac{1}{4}$ "
1853	"	9·85 "	"	$12\frac{4}{7}$ "
1854	"	9·47 "	"	$12\frac{3}{4}$ "
1855	"	8·74 "	"	$12\frac{1}{4}$ "

The following statement, showing the quantities of cotton imported into Great Britain, and the countries whence imported, from 1840 to 1850, is given to illustrate the statement exhibiting the same facts from 1851 to 1855, already presented (III) in answer to the resolution. The figures are derived from a "Statistical Abstract for the United Kingdom in each year from 1840 to 1853, presented to both Houses of Parliament, by command of her Majesty," by Mr. Albany W. Fonblanque, superintendent of the statistical department of the Board of Trade:

Yrs.	Pounds of cotton imported into Great Britain from—						
	U. States.	Brazil.	Egypt.	East Indies.	W. Indies.	Elsewhere.	All countries
1840.	487,856,504	14,779,171	8,324,937	77,011,889	866,157	3,649,402	592,448,010
1841.	358,240,964	16,671,348	9,097,180	97,338,153	1,533,197	5,061,513	487,992,355
1842.	414,030,779	15,222,828	4.489,017	92,972,609	598,603	4,441,250	531,750,086
1843.	574,738,520	18,675,123	9,674,076	65,769,729	1,260.444	3,135,244	673,193,116
1844.	517,218,622	21,084,744	12,406,327	88,639,776	1,707,194	5,054,641	646,111,304
1845	626,650,412	20,157,633	14,614,699	58,437,426	1,394,447	725,336	721,979,953
1846.	401,949,393	14,746,321	14,278,447	34,540,143	1,201,857	1,140,113	467,856,274
1847.	364,599,291	19,966,922	4,814,268	83,934,614	793,933	593,587	474,707,615
1848.	600,247,488	19,971,378	7,231,861	84,101,961	640,437	827,036	713,020,161
1849.	634,504,050	30,738,133	17,369,843	70,838,515	944,307	1,074,164	755,469,012
1850.	493,153,112	30,299,982	18,931,414	118,872,742	228,913	2,090,698	663,576,861

The following table will show the quantities of cotton imported into Great Britain in 1850 and 1851, distinguishing that from foreign countries, and that from the possessions of Great Britain:

* At Manchester.

Pounds of cotton imported into Great Britain.

From foreign countries:	In 1850.	In 1851.
United States, . . .	493,153,112	596,638,962
Brazil, ...	30,299,982	19,339,104
Turkey, Syria, and Egypt, .	18,909,748	15,766,325
Other foreign countries, .	1,619,051	2,141,617
Total from foreign countries,	543,981,893	633,886,008
British possessions:		
East Indies, ...	118,872,742	122,626,976
British West Indies and British Guiana, ...	228,913	446,529
Other British possessions, .	493,313	420,236
Total from British possessions,	119,594,968	123,493,741
Total from foreign countries,	543,981,893	633,886,008
Total of cotton imported, .	663,576,861	757,379,749

Tabular comparative statement showing the declared value of cotton manufacturers of all kinds and cotton yarns exported from Great Britain from 1840 *to* 1850, *both inclusive.*

1840	Manufactures,	$87,836,550	Yarns,	$35,506,540
1841	"	81,162,550	"	36,334,840
1842	"	69,539,420	"	38,857,320
1843	"	81,270,000	"	35,969,855
1844	"	94,083,820	"	34,942,920
1845	"	95,780,480	"	34,816,175
1846	"	88,588,890	"	39,410,240
1847	"	86,876,225	"	29,789,900
1848	"	83,766,845	"	29,639,155
1849	"	100,355,230	"	33,520,445
1850	"	109,368,485	"	31,918,520

FRANCE.

Cotton constitutes, in value, more than two-thirds of the domestic exports of the United States to France. By virtue of the treaty of 1822, it is imported, like all other "articles of the growth, produce, or manufacture of the United States," on the same terms, whether in United States or national vessels; but the importation must be direct, and the origin of the article duly authenticated. A ministerial decree of December 17, 1851, enlarges the provision of the treaty relative to the direct voyage, so far as to extend the equality between the vessels of the two nations when importing cotton, even should the American vessel touch at a British port; but, in that case, the captain is required to exhibit a certificate from the French consul at that port, stating that no commercial transaction there took place.

The French government is directing its efforts to the development and extension of the cotton culture in its colonial province of Algeria. To that end, in December, 1853, an aggregate value of 20,000 francs, in prizes, was offered by the emperor to the most successful cultivator of cotton in that province. The result is announced as most favorable. In December, 1854, the entire sum was divided between three rivals, whose merits were judged equal—two of them being French colonists, and one an Arab—a gold medal to each being also awarded. To the meritorious of the second rank, a silver medal to each was presented. The amount produced in 1854 was 180,552 lbs.

Next to Great Britain, France is the largest importer of American cotton; and what Liverpool is to the former, Havre is to the latter. At those two points the importations are concentrated, and thence distributed to the different markets of either empire or reëxported to foreign countries.

The reëxportations of France are chiefly to Switzerland by railway; after which country, in this trade, come Sardinia and Holland; smaller quantities being sent, also, to Spain, the Zollverein and other countries.

Next to the United States, France derives her supplies of cotton from the Levant; and the third place is held by South America.

These facts are illustrated by the following statements, made up from the "*Tableau General du Commerce de la France,*" for the years designated: [The quantities are given in kilogrammes, each kilogramme being equal to about $2\frac{1}{5}$ lbs. Kilogrammes multiplied by 9 and divided by 4 will give pounds.]

Tabular comparative statement showing the quantities of cotton imported into France, and the countries whence imported, for a period of three years, from 1852 *to* 1855, *both inclusive.*

Kilogrammes of cotton imported into France, in the years—

Countries whence imported.	1852.	1853.	1853.
United States, . . .	76,104,454	79·381,735	77,746,470
Egypt, ...	4,382,575	4,831,872	3,601,327
Turkey, ...	1,027,836	1,371,239	375,834
England, ...	938,907	890,322	1,547,994
Belgium, ...	231,074	603,449	375,350
Brazil, ...	413,563	280,813	127,912
Peru, ...	158,716	233,838	239,688
Venezuela, ...	315,953	169,686	68,064
Hayti, ...	75,697	104,510	77,165
East Indies, ...	47,955	6,674	188,649
Elsewhere, ...	393,091	191,029	206,569
Aggregate, . . .	83,989,822	88,065,167	84,555,022

Tabular comparative statement, showing the quantities of cotton exported by France to all countries, respectively, for a period of three years, from 1852 *to* 1854, *both inclusive, (The quantities are given in kilogrammes, as in the preceding table.)*

Kilogrammes of cotton exported from France in the years—

Countries to which exported.	1852.	1853.	1854.
Switzerland, ...	7,029,667	7,929,099	6,657,003
Netherlands, ...	1,709,004	857,982	688,308
Sardinia, ...	1,554,395	661,924	492,374
Zollverein, ...	193,408	158,637	388,974
Hanse Towns, ...	110,554	182,581	19,264
Austria, ...	17.585	138,636	103,885
England, ...	1,149,966	319,820	77,008
Belgium, ...	75,711	123,061	63,704
Spain, ...	213,863	51,179	53,825
Tuscany, ...	48,915	18,438	1,720
Elsewhere, ...	74,018	30,483	6,493
Aggregate, ...	12,177,086	9,571,840	8,552,558

Comparative tabular statement showing the quantities of cotton consumed in France, and the countries whence imported, for a period of three years, from 1852 *to* 1854, *both inclusive.*

Kilogrammes of cotton consumed in France in the years—

Countries whence imported..	1852.	1853.	1854.
United States, . . .	66,740,104	70,220,752	67,452,503
Egypt, ...	2,754,662	2,401,497	2,318.665
Turkey, ...	979,313	744,331	571,511
England, ...	3,966	8,442	170,664
Belgium, ...	231,074	561,066	395,176
Brazil, ...	432,899	265,450	105,861
Peru, ...	144,134	219,077	254.414
Venezuela, . . .	206,538	161,502	55,263
Hayti, ...	47,860	70,530	57,290
East Indies, . . .	296,953	263,374	71,517
Elsewhere, ...	231,448	175,237	141,131
Aggregate, . . .	72,068,991	75,091,258	71,593,995

Tabular comparative statement showing the quantities of cotton which passed in transit through France, with the countries whence it came and whither it went, respectively, distinguishing the quantities to and from each, for the period of three years, from 1852 *to* 1854, *both inclusive.*

Years.	Countries whence.	Quantities. *Kilos.*	Countries whither.	Quantities. *Kilos.*
1852,	. United States,	5,060,457	Switzerland,	7,027,627
"	. England, . .	1,255,630	Sardinia, . .	364,315
"	. Egypt, . . .	1,025,128	Zollverein, .	196,979
"	. Elsewhere, .	266,319	Elsewhere, .	18,613
	Aggregate, .	7,607,534	Aggregate, .	7,607,534
1853,	. United States,	4,800,328	Switzerland,	7,006,914
"	. England, . .	761,193	Sardinia, . .	197,115
"	. Egypt, . . .	1,822,372	Zollverein, .	192,779
"	. Elsewhere, .	92,178	Belgium, . .	79,263
	Aggregate, .	7,476,071	Aggregate, .	7,476,071
1854,	. United States,	4,623,826	Switzerland,	6,601,925
"	. England, . .	1,402,372	Sardinia, . .	265,387
"	. Egypt, . . .	884,750	Zollverein, .	373,550
"	. Elsewhere, .	386,693	Elsewhere, .	56,779
	Aggregate, .	7,297,641	Aggregate, .	7,297,641

SPAIN.

This kingdom takes from the United States about four-fifths of all her cotton; the quantity, during the last five years, reaching an average of thirty-four million pounds per annum, and showing an increase on the five years immediately preceding. Next to the United States, Spain imports cotton from Brazil, while her West India possessions hold a third rank in the trade.

HANSE TOWNS.

The states of Germany are supplied with the cotton consumed in their factories, chiefly through the Hanseatic cities, Hamburg, and Bremen. Bremen sent to the Zollverein, in 1853, cotton, imported direct from the United States, to the value of $984,772 14; and to Austria to the value of $156,153 21. The factories of Prussia and Saxony are numerous, and import not only the raw material from these cities, but also large quantities of yarns. The number of spindles in operation in the states composing the Zollverein, is estimated at upwards of 1,000,000. This is, doubtless, an under estimate, as the industrial enterprise of the Zollverein has made rapid progress since the date of the official document from which these figures are derived. The export of cotton tissues from the Zollverein, in 1853, amounted in value to $2,394,497 34, of which amount $2,075,299 68 in value, came from the factories of Saxony.

The Hanse Towns, from geographical position, are, and must always continue to be, the great marts from which raw material of all descriptions will be supplied to the states of the Germanic commercial union. Hence, exports of American cotton and tobacco to these points are heavy, and constantly increasing. These commercial cities receive their supplies of raw cotton not only from the United States, in direct trade, as well as from Brazil and other countries of South America, but also in the indirect trade from English ports and other *entrepôts* of Europe. In 1855, the Zollverein sent through the Hanse ports to the United States, cotton fabrics to the value of more than a million and a half of dollars, in return for the raw material.

BELGIUM.

Most of the cotton imported into Belgium, is from the United States, and is consumed by her own factories at Ghent, Liege, Antwerp, Malines, (Mechlin,) &c., &c., which are said to employ a capital of twelve millions dollars, and more than 122,000 operatives, and to turn out an annual value of seventeen millions dollars in fabrics, which are in high repute.

The "*Tableau General*" of Belgium, for 1854, gives the importation of cotton into that kingdom, as follows:

Statement exhibiting the quantities of cotton imported into Belgium, in 1854, in pounds:

From United States,	15,329,266
From England,	14,208,765
From Holland,	2,733,259
From France,	368,516
From Hanse Towns,	79,668
From Hayti,	73,055
From Brazil,	19,991
From other countries,	30,594
Total,	32,833,114

Of the above total, 25,783,292 lbs. was consumed, and 7,049,823 lbs. exported.

The quantity imported by land and rivers, was 3,104,851 lbs.; by sea, 29,729,263 lbs.

Of the quantity exported, 6,959,965 lbs. was by land and rivers, and 89,858 lbs. by sea—

Prussia receiving (by land and rivers) .	5,628,186
France " " " .	842,881
Holland " " " . .	488,898
And all other countries receiving (by sea)	134,118
Total by land and rivers and by sea, .	7,094,083*

The cotton thus exported, was imported as follows:

	Pounds.
From United States, ...	5,529,537
From England, ...	1,488,582
From Holland, ...	70,965
From France, ...	4,999
Total, ...	7,094,083

The average annual amount of duties derived by Belgium from cotton, for the five years ending with 1854, was upwards of $40,000; and, in the latter year, it ranked the thirteenth among articles imported, in this regard. . The duty, under the law of January 31, 1852, was 1 franc 70 centimes per 220 lbs. By the law of April 12, 1854, cotton became free.

In 1854, Belgium exported cotton fabrics in value as follows:

Value of total exports of cotton fabrics, .	$4,701,572
Value of total exports of Belgian manufacture,	2,632,586
Value of total foreign manufactures reëxported,	$2,068,986

SARDINIA.

Sardinia imports, on an average, some four or five millions pounds of cotton each year from England and France, and

* This is an excess of 44,260 lbs. over the amount given above as exported, that quantity having been entered for consumption but subsequently withdrawn.

about the same quantity from the United States; although, in 1855, the importation from the latter country suddenly rose from 1,645,372 lbs. the preceding year, to 14,777,765 lbs.! There seems no sufficient reason why American vessels should not convey the whole quantity required by Sardinia, directly to Genoa, as well as for English or French vessels to carry thither a portion of American cargoes landed at Liverpool or Havre. A similar remark is applicable to the other ports of Italy, and to those of Austria on the Adriatic; and the enterprise of establishing lines of ocean steamers between ports of the United States and those of the Mediterranean, will, if successful, tend greatly to encourage, if not to secure, such direct importation.

SWITZERLAND.

Four-fifths of all the cotton consumed by the factories of Switzerland, is estimated to be imported at Havre, whence it passes through France by railway, being burdened with heavy charges in the transit. In 1833, the quantity thus received amounted to nearly 6,000,000 lbs. In 1843, it had reached nearly 17,000,000 lbs. The entire receipt of cotton in 1843, was 22,000,000 lbs. In 1851, it amounted to 27,035,725 lbs., of which 13,729,320 lbs. were from the United States. In 1852, Switzerland received through France, 15,816,775 lbs.; in 1853, 15,815,473 lbs.; and in 1854, 14,978,257 lbs., according to the "*Tableau General*" of France, for those years.

Imports from the United States into Switzerland, are made, for the most part, through the customs frontiers of Berne, Basle, Soleure, and Argovie, bordering on France and the southern part of Germany.

A severe restriction on the importation of cotton, and also of tobacco, to Switzerland, as well as on the reception by the United States of Swiss wares and manufactures in return, is the vexatious and expensive transitage, especially through France. The oppression of this system may be inferred from the fact that the annual average aggregate value of merchandize on which transit tolls are paid, proceeding from Switzerland, is (1853) nearly thirty millions dollars; and the value of that proceeding to that republic, more than half as much.

Switzerland sent, *in transitu* to France, cotton tissues to the value of nearly three millions dollars in 1852; and to the value of nearly four millions, in 1853. By the French tariff, such fabrics are excluded from France for consumption. Since 1845, Switzerland is stated, officially, to have quite superseded in the markets of Germany and Austria, the yarns of Great Britain. In 1830, that republic had in operation 400,000 spindles; in 1840, 750,000; and in 1850, 950,000; the number having more than doubled in twenty years.

According to Swiss official custom-house reports, that republic received cotton from the United States as follows, the years specified:

	Pounds.
1850,	15,942,740
1851,	13,729,320
1852,	19,065,200
1853,	18,441,830

In return, cotton stuffs, as follows, were sent to the United States:

1850,	3,226,300
1851,	3,509,660
1852,	4,077,920
1853,	5,265,150

In 1855, Switzerland returned to the United States, in exchange for raw cotton, the same article manufactured, to the value of $212,700.

RUSSIA.

Before the breaking out of the late war, the manfacture of cotton in the Russian empire was progressing with extraordinary activity. The number of spindles exceeded 350,000, producing, annually, upwards of 10,800,000 lbs. of cotton yarns. The barter trade with the Chinese, at Kiachta, stimulates this branch of manufacture in Russia, as the article of cotton velvets constitutes the leading staple of exchange, at that point, for the teas and other merchandise of China. In former years this article was supplied almost exclusively by Great Britain; but the Chinese prefer the Russian manufacture, and hence the steady progress of that branch of industry. Thus the annually increasing importations of the raw material, and consequent diminution in the quantities of cotton yarns imported is accounted for. Were raw cotton admitted, as in England, free of duty, the United States would, most probably, supply, in the direct trade, the whole quantity consumed in that empire. As it is, the commercial reforms in Russia, already announced officially, and now in progress, comprehending, as they do, the establishment of American houses at St. Petersburg, must necessarily tend to that result.

There are, at present, in Russia, or there were, previously to the war, 495 cotton factories, employing 112,427 operatives, and producing, annually, 40,907,736 lbs. of yarns, and corresponding amounts of textiles.

SWEDEN.

The importation of cotton in 1851, according to Swedish official authorities, amounted to 7,989,428 lbs., against 1,832,431 lbs. in 1841, and 794,434 lbs. in 1831. In 1843 these authorities show an importation of 2,600,000 lbs., against 9,888,572 lbs. in 1853; which latter amount exceeded that of the importation of 1852 by 1,247,041 lbs., and that of 1850 by more than 5,200,000 lbs. being the largest of any preceding year. In 1848 the amount was 8,074,020 lbs.

The value of cotton manufactures exported from Sweden in 1850 was $46,000, against $7,500, only, in 1851.

PORTUGAL.

This kingdom imported 1,911,451 lbs. of cotton in 1855, of which quantity 144,006 lbs. were exported from the United States, and the residue from Brazil. In 1853–'54, according to Brazilian official reports, Portugal received thence 2,673,766 lbs. of cotton. Her imports of yarn in 1855 were 1,213,157 lbs., valued at $171,817 07, and paying an aggregate of duties of $61,142 84.

BRAZIL.

The exportations of cotton from Brazil in 1843–'44 and 1853–'54 are stated, by Brazilian official authorities, as follows:

1853–'54, ...	28,420,320 lbs.
1843–'44, ...	26,056,160 "
Increase in ten years, . . .	2,364,160

In 1852–'53, the exportation amounted to 31,933,050 lbs., of which quantity Great Britain received 26,881,201 lbs., Spain 2,291,578 lbs., Portugal 1,896,286 lbs., and France 889,048 lbs.

Of the total exportations in 1853–'54, Great Britain received 22,575,122 lbs., Spain 2,351,279 lbs., Portugal 2,673,766 lbs., and France 543,611 lbs.

Exports from Brazil to England began in 1781.

There are insuperable drawbacks to the extension of the cotton culture in Brazil, among which may be reckoned the ravages of insects, the peculiarities of the climate, and the expense and difficulties attendant upon its transmission from the interior to the coast. It has long since been ascertained in Brazil that the cotton plant will not flourish near to the sea, and the plantations have, in consequence, receded further inland, as well to avoid this difficulty as to seek new and fresh lands. Pernambuco is the principal cotton-growing province of Brazil. The exports from that province were, according to Brazilian authorities, in—

	BALES, 160 LBS. EA.		BALES, 160 LBS. EA.
1828, . . .	70,785	1840, . . .	35,849
1830, . . .	61,151	1842, . . .	21,357
1835, . . .	52,142	1845, . . .	26,562

EGYPT.

The cotton culture in Egypt commenced in 1818, and exportation to England in 1823.

The comparative tabular statement subjoined, derived from Egyptian sources, showing the quantities exported at the port of Alexandria, and the countries to which exported, respectively, for a period of three years, from 1853 to 1855, both inclusive, would indicate an increase in the culture by no means rapid in successive years:

Pounds of Cotton exported to—

YEARS.	GREAT BRITAIN.	FRANCE.	AUSTRIA.	ELSEWHERE.	ALL COUNTRIES.
1853,	26,439,900	10,726,500	6,321,000	397,800	43,885,200
1854,	24,938,700	7,454,100	10,165,200	988,500	43,546,500
1855,	33,980,100	9,451,200	12,774,900	668,100	56,874,300
Agg.,	85,358,700	27,631,800	29,261,100	2,054,400	144,306,000
Av.,	28,452,900	9,210,600	9,753,700	684,800	48,102,000

If to the aggregate exported be added from five to six million pounds worked up in the country, a liberal estimate of the annual amount of the cotton crops of Egypt will have been made. The factories established by Mehemet Ali are, it is stated, going rapidly to ruin. The cotton goods manufactured are coarse "caftas" or soldier's "nizam" uniform. Much cotton is used also, in making up divans, the usual furniture in Egypt. The Egyptian bale is estimated at Alexandria at 300 lbs. The United States consul-general at that port, in a dispatch dated the 1st instant, from which are derived the above facts, says: "The new crop is now coming in, and is supposed to be a little above the average."

CAPACITY OF THE COTTON BALE.

The commercial standard of quantity in the cotton trade is generally the bale. The weight of the bale, however, is by no means uniform. Indeed, scarcely any weight, measure, or standard of capacity may be considered less so. It varies, from different causes, in different countries, and in different sections of the same country, at different periods, and according to the different kinds or qualities of the article. Improvements in pressing and packing, to diminish expense in bagging and freight, tend constantly to augment the weight of the bale. Thus, in 1790, the United States bale was computed at only 200 lbs. In 1824 the average weight of bales imported into Liverpool was 266 lbs.; but, increasing constantly, twelve years later the average was 319 lbs McCulloch, however, in 1832, considered 300 to 310 lbs. a fair average; and Burns 310. At the same time, the upland cotton bale was estimate at 320 lbs., and the sea island at 280 lbs. According to Pitkins, the Egyptian bale weighed at one time but 90 lbs., though it now weighs more than three times as many. At the same period the Brazilian bale contained 180 lbs., though it now contains but 160 lbs., while the West Indian bale weighed 350 lbs., and the Columbian bale 101 lbs., or the Spanish quintal. According to Burns, the United States bale at Liverpool averaged 345 lbs.; the Brazilian 180 lbs.; the Egyptian 220 lbs.; the West Indian 300 lbs., and the East Indian 330 lbs. At the Lowell factories, in 1831, according to Pitkins, the bale averaged 361 lbs. In 1836 the bale of the Atlantic cotton States was estimated at 300 and 325 lbs., and that of the Gulf States at 400 and 450 lbs. In Liverpool, at the same time, the estimate for the bale of upland or short staple cotton was 321 lbs.; for Orleans and Alabama 402 lbs.; for sea island 322 lbs.; for Brazil 173 lbs.; for Egyptian 218 lbs.; for East Indian 360 lbs., and for West Indian 230 lbs.; while, according to Burns, bales imported into France were computed at only 200 lbs. each. Waterton's "*Manual of Commerce*," a reliable British publication, (1855,) gives the Virginia, Carolina, Georgia, and West Indian bale at 300 to 310 lbs.; that of New Orleans and Alabama at 400 to 500 lbs.; East Indies at 320 to 360 lbs.; Brazil at 160 to 200 lbs. Egyptian at 180 to 280 lbs. Alexander's "Universal Dictionary of Weights and Measures" gives the bale of Alabama, Louisiana and Mississippi at 500 lbs.; that of Georgia at 375 lbs., and that of South Carolina at 362 lbs. At Rio de Janeiro the Brazil bale is estimated at 160 lbs.

Prior to 1855, the United States "Commerce and Navigation" gave exports of cotton in pounds only. They are now given in bales as well as in pounds, the aggregate amount for the year ending June 30, 1855, being 2,303,403 bales, or 1,008,-424,601 lbs.; the bale, accordingly, averaging about 438 lbs. Some bales, however, are evidently much heavier and some much lighter than this. For example, the 210,113,809 lbs. of cotton exported to France give 446 lbs. to each of the

470,293 bales; and the 955,114 lbs. exported to Austria give 492 lbs. to each of the 1,939 bales; while the 7,527,079 lbs. exported to Mexico give only 290 lbs. to each of the 25,917 bales in which they were contained.

The relative average weights and cubical contents of bales of cotton imported into Liverpool in 1852 are thus given:

Descrip. of bales.	Av. wt. in lbs.	Con. in cub. ft.	Descrip. of bales.	Av. wt. in lbs.	Con. in cub. ft.
Mobile, . . .	504	33	East Indian, .	383	15
New Orleans, .	455	32	Egyptian, . .	245	27
Upland, . .	390	27	West Indian, .	212	25
Sea Island, . .	333	35	Brazilian, . .	182	17

These figures show not only the great variety of bales that enter Liverpool, but that the most eligible form of bale is that of the East Indies—double the weight being packed within the same compass as in any other description of bale.

In the great cotton marts of Liverpool and Havre, as in those of New Orleans and Mobile, the article is almost invariably treated of by merchants, brokers, and commercial men, by the bale. Thus, a report on the trade of Liverpool gives the imports of cotton into Great Britain, in 1852, at 2,357,338 bales. The aggregate of cotton imported that year is given in the official report by the Board of Trade, at 929,782,448 lbs.; the bales averaging, accordingly, 395 lbs. each.

The annual commercial "*Revue*" of Havre, gives the number of bales of cotton imported into France, the same year (1852) at 462,300 in round numbers. The "*Tableau General*" gives the imports at 188,917,099 lbs.; the bales averaging, accordingly, about 409 lbs. each.

The following table, compiled from the Havre commercial "*Revue*," (1855,) referred to, shows the quantities of cotton, in bales, imported into France, and the countries whence imported, for a period of five years, from 1851 to 1855, both inclusive:

YEARS.	UNITED STATES.	BRAZIL.	EGYPT.	ELSEWHERE.	ALL COUNTRIES.
1851	295,400	7,700	18,500	38,000	359,600
1852	392,700	6,000	36,700	26,900	462,300
1853	389,000	2,800	33,000	29,200	454,000
1854	403,300	2,000	21,400	16,300	470,000
1855	418,600	2,500	30,700	11,800	463,000

Estimating the bale at 400 lbs., we have the following statement, some of the figures of which, contrasted with those derived from official sources in the statement already given, (III,) present striking discrepancies.

Tabular comparative statement showing the quantities of cotton, in round numbers, imported into France, and the countries whence imported, for a period of five years, from 1851 *to* 1855, *both inclusive, the bale being estimated at* 400 *lbs.*

Pounds of cotton imported into France from—

YEARS.	UNITED STATES.	BRAZIL.	EGYPT.	ELSEWHERE.	ALL COUNTRIES.
1851, . .	118,160,000	3,080,000	7,400,000	15,200,000	143,840,000
1852, .	157,080,000	2,400,000	14,680,000	10,760,000	104,920,000
1853, . .	155,600,000	1,120,000	13,200,000	11,680,000	181,600,000
1854, .	172,120,000	800,000	8,560,000	6,520,000	188,000,000
1855, . .	167,440,000	1,000,000	12,280,000	4,720,000	185,440,000
Aggregate,	770,400,000	8,400,000	56,120,000	48,880,000	803,800,000
Average,	154,080,000	1,680,000	11,224,000	9,776,000	160,760,000

NOTE.—Marked discrepancies are perceived in statements of the same statistical facts, for the same periods, derived from official data of different countries. Although some such discrepancies may be rather apparent than real, and attributable to variations in the terminations of commercial years, while for others various causes, more or less satisfactory, may be assigned, it still remains a vain task to attempt the entire reconciliation of these statistical conflicts.

CHAPTER VIII.

HISTORY OF COTTON, AND THE COTTON GIN.

SECTION I.—BRIEF HISTORY OF COTTON.

COTTON, which administers so bountifully to the wants of civilized as well as to savage man, and to the wealth and economy of the countries producing it, stands foremost among the crops in the United States, both as regards its superior staple and the degree of perfection to which its cultivation has been brought. One or more of its species is found growing wild throughout the torrid zone, whence it has been disseminated and become an important object of culture in several countries adjacent, where its consumption has increased just in proportion to the progress of the arts and civilization. It is mentioned by Herodotus as growing in India, where the natives manufactured it into cloth; by Theophrastus, as a product of Ethiopia; and by Pliny, as growing in Egypt, towards Arabia, and near the borders of the Persian Gulf. Nienhoff, who visited China in 1655, says that it was then cultivated in great abundance in that country, where the seed had been introduced about five hundred years before. Columbus found it in use by the American Indians of Cuba, in 1492; Cortez, by those of Mexico, in 1519; Pizarro and Almagro, by the Incas of Peru, in 1532; and Cabeça de Vaca, by the natives of Texas and California, in 1536.

Of the precise period of the first introduction of the cotton plant into the North American colonies, history is silent. In a pamphlet entitled, "Nova Brittania, offering most excellent fruits by planting in Virginia," published in London in 1609, it is stated that cotton would grow as well in that province as in Italy. It is also stated, on the authority of Beverley, in his "History of Virginia," that Sir Edmund Andros, while governor of the colony, in 1692, "gave particular marks of his favor towards the propagating of cotton; which, since his time, has been much neglected." It further appears that it was cultivated for a long time in the eastern parts of Maryland, Virginia, Carolina, and Georgia, in the garden, though not at all as a planter's crop, for domestic consumption.

In another pamphlet, entitled "A state of the province of Georgia, attested upon oath in the court of Savannah," in 1740, it was averred that "large quantities have been raised, and it is much planted; but the cotton, which in some parts is perennial, dies here in the winter; which, nevertheless, the annual is not inferior to in goodness, but requires more trouble in cleansing from the seed." About the year 1742, M. Dubreuil invented a cotton gin, which created an epoch in the cultivation of this product in Louisiana. During the Revolution, the inhabitants of St. Mary's and Talbot counties, in Maryland, as well as those of Cape May county, New Jersey, raised a sufficient quantity of cotton to meet their wants for the time. It was formerly produced in small quantities, for family use, in the county of Sussex, in Delaware, near the head waters of the Choptank.

The seed of the Sea Island cotton was originally obtained from the Bahama Islands, in about the year 1785; being the kind then known in the West Indies as the "Anguilla cotton." It was first cultivated by Josiah Tatnall and Nicholas Turnbull, on Skidaway Island, near Savannah; and subsequently by James Spaulding and Alexander Bisset, on St. Simon's Island, at the mouth of the Altamaha, and on Jekyl Island by Richard Leake. For many years after its introduction, it was confined to the more elevated parts of these islands, bathed by the saline atmosphere, and surrounded by the sea. Gradually, however, the cotton culture was extended to the lower grounds, and beyond the limits of the islands to the adjacent shores of the continent, into soils containing a mixture of clay; and, lastly, into coarse clays deposited along the great rivers, where they meet the ocean tides.

Previous to 1794—the year after the invention of Whitney's saw-gin—the annual amount of cotton produced in North America was comparatively inconsiderable; but since that period, there is probably nothing recorded in the history of industry, including its manufactures in this country and Europe, that would compare with its subsequent increase.

The earliest record of sending cotton from this country to Europe, is in the table of exports from Charleston in 1747–'48, when seven bags were shipped; another parcel, consisting of 2,000 lbs., was shipped from the same port in 1770; and a third shipment of seventy-one bags was made in 1784, which was seized in England on the ground that America could not produce a quantity so great. In 1792, there were shipped 304 bales; in the first six months of 1796, 150 bales. From an old custom-house book at Wilmington, North Carolina, it appears that in July, 1768, the ship "Amelia" cleared from that port with an assorted cargo, among which were three bags of cotton. In 1796, there were exported from Philadelphia 911,325 lbs.

The amount of cotton exported from the United States in 1791, was 189,316 lbs.; in 1793, 187,600 lbs.; in 1794, 1,601,760 lbs.; in 1795, 6,276,300 lbs.; in 1800, 17,789,803 lbs.; in 1810, 93,261,462 lbs.—[*Patent Office Report for* 1853, *Agricultural Dept., p.* 179.]

From the Southern Agriculturist.

SECTION II.—ON THE COTTON GIN, AND THE INTRODUCTION OF COTTON.

Answers to queries of Hon. W. B. Seabrook of Edisto, S. C., by Thomas Spalding, Esq., of Sapelo, Ga.

SAPELO ISLAND, January 20th, 1844.

DEAR SIR:—Your letter of the 10th instant was received two days ago, and I was gratified at the communication, as I have long wished to be personally acquainted with some of the gentlemen of your immediate district; your pursuits, your habits, and your opinions, appearing to be in accordance with my own; and nothing but the continued pressure of a painful disease, of now ten years' standing, has prevented me carrying out my design, by visiting you.

I will now proceed to answer your queries, in the order in which they are placed; only begging you to remember, if you notice any indistinctness in my answers, that I have only a few days since recovered from a very severe illness which prostrated both body and mind.

1st. Eve's Gin was invented by Joseph Eve, who lately died at Augusta, somewhere about the year 1790, in the Bahama Islands, where Mr. Eve then resided.

Mr. Eve was the son of a Loyalist from Pennsylvania, who had been the friend of Franklin; and Joseph Eve was himself qualified to have been the associate and companion of Franklin, or any other; the most enlightened man I have ever known. His gin consists of two pair of rollers, more than three feet long, placed the one set over the other, upon a solid frame that stands upon the floor, inclined at an angle of about thirty degrees—so that the feeder may the more easily throw the cotton in the seed by the handful upon a wire grating that projects two inches in advance of the rollers, just below them; between these projecting wires, the feeding-boards, with strong iron, or in preference, brass teeth pass, lifting the cotton from the wire grating, and offering it to the revolving rollers. The feeders should make one revolution to every four revolutions of the rollers. The rollers are carried forward by wheels supported over the gin, and upon the axle or shaft of these rollers; at the center there is a crank similar to a saw-mill crank, the diameter of whose revolvement is as one to four of the diameter of the wheels, carrying by bands the rollers.

It is the crimping produced by the teeth and the wire grating, which has served as a cause for carping by the cotton-buyers, and which has gradually led to the disuse of these gins, the only gin efficient for the cleaning of long cotton, which has ever been used in this or any other country. With Mr. Eve's gin, as originally sent to this country from the Bahamas, the rollers were five-eighths of an inch in diameter, made of stopper wood, a very hard and tough wood, and they were graduated to make four hundred and eighty to five hundred revolutions per minute, depending of course upon the gait of the horses or mules, within these limits. Soon after Mr. Eve first sent his gins to Georgia, some of his own workmen followed them, and began to make them on their own account. To show as much change as possible in the gins, beside the other alterations, they increased the size of the rollers, they increased the size of the rollers to three-fourths of an inch, and increased its velocity to six hundred times in the minute. These two changes, while they greatly increased the quantity ginned, very much injured the appearance of the ginned cotton. Mr. Eve had expected and guaranteed to the purchasers of his gins when well attended, in fine weather, from two hundred and fifty to three hundred pounds of cotton in the day. I have known these altered gins do sometimes six hundred, but the injury was greater than the increased quantity warranted, add to which the quicker movement of the feeder made the more impression upon the cotton passing from the feeder to the roller.

2d. The first bale of Sea Island cotton that was ever produced in Georgia, was grown by Alexander Bisset, Esq., of St. Simon's Island, and I think in the year 1778. In the winter of 1785 and '86, I know of three parcels of cotton seed being sent from the Bahamas, by gentlemen of rank there, to their friends in Georgia; Col. Kelsall sent to my father a small box of cotton seed; the surveyor-general of the Bahamas, Col. Tatnall, sent to his son, afterwards Governor Tatnall of Georgia, a parcel of cotton seed; Alexander Bissett's father, who was commissary-general to the Southern British Army, sent a box of cotton seed to his son, in the year 1786; this cotton gave no fruit, but the winter being moderate and the land new and warm, both my father and Mr. Bissett had seed from the ratoon, and the plant became acclimatized. In 1788, Mr. Bissett and my father extended the growth, but upon my memory it rests, that Mr. Bissett was the first that found the means of separating the seed from the cotton, by the simple process of a bench upon which rose a frame supporting two short rollers, revolving in opposite directions, and each turned by a black boy or girl, and giving as the result of the day's work five lbs. of clean cotton. What disposition Mr. Bissett made of his cotton I know not, but as he was a sensible man, and his father had returned to England, I think it more than probable that he shipped it there.

3d. When cotton was first grown, it was planted on the flat land at five —— apart; it was quite too thin, and although the plant grew generally well, the product rarely reached one hundred pounds per acre, and at four acres to the hand, gave about four hundred to the labor.

My father died in the year 1794, leaving me some property at St. Simon's Island. A gentleman, who had been his friend, came, for his health, to spend the winter of that year in Georgia; he gave his advice freely to all he saw that were growing cotton. I was young, he had been the friend of my father, I listened to his advice, left eight or ten plants where one had grown, and made off a small field of sixty acres, 350 lbs. to the acre. The revolution was accomplished, and the crop greatly increased.

4th. No manure was used for many years in the culture of cotton, persons depending upon the in-field and the out, or the alternate cultivation of the field, which was soon found necessary. The first suggestion of manure upon a large scale to cotton, came from Col. Shubrick, of South Carolina, who recommended, in some essays in the papers, the application of the drifted reck that is thrown up by the tides. After the hurricane of 1804, I bestowed a great deal of labor in spreading this reck between my cotton rows, over several hundred acres. Whether the sea had left too much salt, or whether there was too much in the material itself, I know not, but I neither then, nor afterwards, experienced much benefit from the application.

5th. The plough was but little used for any purpose at St. Simon's. It takes many years before the palmetto, and the collateral roots of the live oak, make hammock land free to the plough. Major Butler did use the plough, with mules, for both purposes.

6th. The cotton was generally worked four times. We soon found that our working should cease as soon as the rains became heavy, say at the middle or end of July.

7th. The ridges were renewed every year, or every other

year, whenever the field was planted in cotton. They were originally low, and rather small they were increased in height and breadth, according to the different opinions of men.

From the year 1798 to 1802, the St. Simon's cultivation had assumed a regular form, and was in my opinion good; twenty-one rows to the 105 feet, the ridge occupying the entire space, large, but full and flat upon the top. The cotton seed drilled, and the plants thinned, from six to ten inches apart, dependant upon the expected growth of the plant. Major Butler, and Messrs. Couper and Hamilton, who cultivated extensively near me, were in the habit of topping the cotton in August, to retain, as they supposed, its fruit. I was in the habit of taking off the top of the plant, when the cotton was from 15 to 18 inches high, to make it branch and give a better head.

Twenty years ago, upon purchasing some river land opposite to Savannah, I adopted permanent ridges, planting a row of corn, and a row of cotton, alternately. These ridges had stood nine years when my son sold the plantation, giving, as I think, the best cotton and the best corn crops in Chatham county. And this course I consider the nearest approach to Flemish husbandry I have known in Georgia; because, although the corn and the cotton changed alternately from ridge to ridge, the entire field was kept in full culture, preventing the growth of grass and noxious weeds.

8th. Accounts were kept in pounds, shillings, and pence, in those times. Cotton brought at first 14*d.* sterling, but rose gradually, in about four or five years, to two shillings; at which it stood until the unfortunate dabbling with commerce commenced in the year 1806. The first non-importation act passed in that year, and none more active in its adoption than our southern men. There were but five men, south of the Potomac, who voted against it: Randolph, J. M. Garnett, Thomson of Virginia, Standard of North Carolina, and myself from

Georgia. From that day to this, the agriculture and commerce of the country has been at the mercy of speculators.

9th. Care was taken, for many years, as much as possible to separate the seed carrying any fur, from the black seed intended for planting.

10th. The St. Simon's cotton stood first, and Major Butler's and my own first among them. From the character of the tradesmen attending his gins, or the greater strictness of his manager, his cotton soon took a preference which it preserved for some years. The staple of the St. Simon's cotton was thought better than any other; the putting up of Maj. Butler's cotton placed it at the hands of others.

11th. The bags were packed as now, with the pestle. I never knew the screw used for long staple cotton but at Mr, Hamilton's plantation, and it was soon given up.

12th. The green seed cotton was for some years packed with the pestle; in fact I remember to have heard objections made to the screw, and square bales, at their first introduction.

13th. Negroes were worked in task-work, in Carolina and Georgia, upon the sea-coast, from my earliest recollection. The task in listing, the fields being previously cleared up and the remains of the former year burned off, was half an acre; the laborer was required to ridge afterwards, when carefully done, three-eighths of an acre; and in hoeing, half an acre was the task, depending, however, much upon the season and the condition of the field.

14th. The bags generally were expected to weigh 300 lbs. Major Butler's were $4\frac{1}{4}$ yards, and contained 260 lbs.

15th. The rust did not appear for some years in our fields, and when it did, I attributed it to listing in green vegetable matter in the latter summer, for the next year's crop; or coarse vegetable matter in the winter, instead of burning off, which left a light top-dressing of ashes. The caterpillars made their appearance, I think, for the first time, in 1793, and destroyed the crop. I remember Major Butler made but eighteen bales of cotton from 400 acres. There was also a red bug, a winged insect with a long proboscis, with which it pierced the green pods, extracting the juices of the seed and leaving the pod blighted and hard, and the cotton stained of a deep yellow or red color. In new lands this insect was very destructive, as it had been in the Bahamas; and as it found protection against the cold in the bark and roots of the trees, it was apt to remain for years, injúring the quality and reducing the quantity of the cotton.

16th. The caterpillar was first seen to do injury, as I think, in 1793: the injury was unusual, the destruction complete, so as scarcely to leave seed. The destructive caterpillar is not the same that feeds upon the indigo; the green caterpillar I have frequently known to riddle the leaves to a great extent, without great ultimate injury; but it is the black and yellow striped caterpillar that in a few days, say from four to five, will spread over hundreds of acres, not leaving a green leaf, and finally nothing but the full grown pods, which they sometimes break and injure.

17th. The black seed cotton had been shipped for several years, before they began to grow in the interior the green seed for sale.

I remain, dear sir, respectfully, your ob't. serv't,

THOMAS SPALDING.

SECTION III.—NATHAN LYONS.

Mr. Editor:—Can you or any of your correspondents give the public any account of Nathan Lyons? His name seems to be almost forgotten; yet, if tradition can be relied on, few men have done more for the South—since it is to him, it seems, that cotton planters are indebted for that indispensable machine, the cotton gin, now in such general use. It is just cause of reproach to any people, to forget or to refuse due honors to their own benefactors; and it is for the purpose of retrieving, in some sort, this country from such a reproach, that I would now solicit for publication in the *Soil of the South*, a brief memoir of Nathan Lyons, a man whose inventive genius, it is asserted, first contrived the circular saw for separating the seed from the cotton wool.

Eli Whitney, of Connecticut, doubtless developed the first idea of a machine for ginning cotton by the use of a single revolving cylinder armed with iron points or teeth, acting in connection with fixed bars, and a bush to extricate the cotton wool from the teeth of the revolving cylinder. For this invention he obtained a patent right of exclusive use and sale, and erected one or more machines at Augusta about the close of the last century, or the first of this. A glimpse of this machine, it is said, suggested to the quick mind of Lyons, the substitution of a circular saw, for the wire hooks or card-teeth contrivance of Whitney. Although the invention of Lyons had a practical value incomparably greater than that of Whitney, it was never patented—the inventor contenting himself with such profits as might accrue from the manufacture and sale of saw-gins at his own shop. Though Whitney is admitted to be fairly entitled to the honor of originality, except a grant of money from the government of South Carolina, it is not believed that he ever derived any considerable pecuniary profits from his patent. He instituted suits against many persons in this State who were using Lyons' saw-gin, but it is not known that he recovered in a single instance. The ablest counsel could not make our juries believe that Whitney's model was the same piece of machinery with Lyons' saw-gin. Since the days of Lyons, Griswold, Taylor, Reid, Oglesby, and other machinists of this State, have, by various improvements, brought the saw-gin to great perfection; but whilst conceding all due praise to them, let us not be guilty of slighting original inventors, especially those whose mental achievements have contributed so much to enrich and aggrandize our own Georgia. Respectfully,

Blakeley, Ga., May, 10, 1852. J. C.

[We approve the suggestions of our correspondent from Blakeley, and hope that they may be the means of eliciting the desired information. There are doubtless some persons residing in Middle Georgia, perhaps in Putnam or Hancock counties, who could furnish us with such a memoir. The importance of the invention, or the honor due to the name of the man who made it, is enhanced as it grows older; and whilst those who were the contemporaries of such a man may not properly estimate his claims to distinction, it is due to the history of the times, as well as to the fame of the man, that we should properly record the facts in the case, and give "honor to whom honor is due." Monuments rise, and history teems with eulogies upon the valorous chief of the battle field, while the name of the humble, unpretending mechanic, who may have originated improvements which swelled the wealth of the nation, saved labor, and ameliorated the condition of the whole human family, may have failed to descend to posterity, or find a place in the history of the country. It may not have been a matter of any great importance when Georgia was the little and obscure member of the old family of thirteen, to have known who lived or figured in that day. But now, when her cognomen is, "The Empire State of the South," we want to go back, (as with our great men,) to her school boy days, note the indices of those times, and see what and who hath helped to build up this great name.]

SECTION IV.—ORIGIN OF THE COTTON GIN.

Mr. Editor :—In a last year's number of the *Soil of the South,* some one has written an article respecting the invention of the cotton saw-gin, and seems to think that Nathan Lyons was the inventor. My father, who settled in this place before the Revolution, (and this is the oldest town in Georgia,) has often told me that Jesse Bull, the father of Col. O. A. Bull, now of La Grange, and Charles M. Lin, and Lyons, were all interested in the making of the first gin. Mr. Lin is the only survivor of the trio. He is a poor man; resides in or near Oxford, Georgia. Being on a visit here I had a conversation with him a few days since and gathered from him the following particulars. The first cotton gin was put in operation at the mill on Brier Creek run by water, nine miles below this. This gin was said to be invented by Whitney—it was not made of saws—but with teeth, something like the cotton card—it was kept concealed—the man who tended it was ordered to let no man in to see it; women, who were many of them very anxious to see it, were admitted—at the same time Mr. Bull, being a man of great mechanical genius, was closely engaged trying to construct a machine for separating the seed from the lint. Lyons—Ned Lyons—was at work with him, and proposed to go in disguise and see the gin then in operation—and did—dressed himself in women's apparel—went in and examined it—this fact is corroborated by Mr. Hobson Bacon, whose brother was hired to tend the gin—and he, the brother, was taken sick, and my uncle Hobson Bacon took his place for a few days--during which time many women were in to see the cotton gin. Soon after the first saw gin came out from Jesse Bull's shop, was put up in a house on Broad street in this place—and was run by this same man, Chas. M. Lin. These I believe to be facts—I remember the house myself. After this, Miller and Whitney, the patentees of the Whitney gin, brought an action against the inventors of the saw-gin, and after about nine years' litigation, got judgment for a few dollars. Mr. Lin thinks he had to pay one hundred and fifty dollars for running the gin, but there was injustice in the whole proceeding. Miller & Whitney feed every attorney that could be employed, and the case was carried up by Lyons to the Federal court and there he, Lin, was lost sight of. Bull turned into South Carolina, and after being absent for months, at sundry times, would always return with heavy bags of silver. Another circumstance connected with this history is, that Dr. De Amford, a German, living in this place, was called to give his testimony in the court, and he swore that the gin which Whitney professed to be the inventor of, was invented by a surgeon of Germany, for the preparing of lint for the army.

These are all the facts I can gather from these two old men. I have hastily thrown them together in bad taste. If you think them worth a place in your paper, you can use them. Very respectfully,

Wrightsboro', Ga., Jan., 1853. THOMAS H. WHITE.

SECTION V.—STATISTICS OF COTTON.

THE following brief items of the history of cotton for about a hundred years—1730 to 1836—will be read and referred to with interest:

1730. Mr. Wyatt spins the first cotton yarn in England by machinery.

1735. The Dutch first export cotton from Surinam.

1742. First mill for spinning cotton erected at Birmingham, moved by mules or horses; but not successful in operations.

1749. The first shuttle generally used in England.

1756. Cotton velvets and quiltings in England for the first time.

1761. Arkwright obtained the first patent for the spinning frame, which he further improved.

1768. The stocking frame applied by Hammond to making of lace.

1773. A bill passed to prevent the export of machinery used in cotton factories.

1779. Mule spinning invented by Hargrave.

1782. First import of raw cotton from Brazil into England.

1782. Watt took out his patent for the steam engine.

1783. A bounty granted in England on the export of certain cotton goods.

1785. Power looms invented by Dr. Cartwright. Steam engines used in cotton factories.

1785. Cotton imported into England from the United States.

1786. Bleaching first performed by agency of the oxy-muriatic acid.

1787. First machinery to spin cotton put in operation in France.

1789. Sea island cotton first planted in the United States; and upland cotton first cultivated for use and export about this time.

1790. Slater, an Englishman, built the first American cotton factory at Pawtucket, R. I.

1792. Eli Whitney, an American, invents the cotton gin, which he patents.

1798. First mill and machinery for cotton erected in Switzerland.

1799. Spinning by machinery introduced into Saxony this year.

1803. First cotton factory built in New Hampshire.

1805. Power looms successfully and widely introduced into England.

1807. The revolution in Spanish America begins to furnish new markets for cotton manufactures.

1810. Digest of cotton manufactures in the United States by Mr. Gallatin, and another by Mr. Tenche Cox, of Philadelphia.

1811. Machinery to make bobbin lace patented by John Bern.

1813. The Indian trade more free, and more British manufactures sent thither.

1814. The power-loom first introduced into the United States; first at Waltham.

1818. Average price of cotton 34 cents—higher than since 1810. New method of preparing sewing cotton by Mr. Holt.

1819. Extraordinary price for Alabama cotton lands.

1820. Steam power first applied, with success, extensively to lace manufactures.

1822. First cotton factory in Lowell erected.

1823. First export of raw cotton from Egypt into Great Britain.

1825. New Orleans cotton at from 23 to 25 cents per pound.

1826. Self-acting mule spinner patented in England by Roberts.

1827. American cotton manufactures first exported to any considerable extent.

1829. Highest duty in the United States on foreign cotton manufactures.

1830. About this time Mr. Dyer introduced a machine from the United States into England, for the purpose of making cards.

1832. Duty on cotton goods imported into the United

States reduced; and in England it is forbid to employ minors in cotton mills to work more than ten hours per day, or nine hours on Saturday; in consequence they work at something else.

1834. Cotton at 17 cents.

1835. Extensive purchases made of cotton lands by speculators and others.

1836. Cotton from 18 to 20 cents.

SECTION VI.—COTTON GIN AND PACKING SCREW.

MESSRS. EDITORS:—One of your correspondents has discovered that cast-iron screws are a desideratum in packing cotton, and refers to Mr. Finley, of Macon, for the cost of them. This is no new discovery. The first screw employed in packing cotton was probably made of iron. About the year seventeen hundred and ninety-five, a gentleman from Baltimore—the father of Judge Bull, of La Grange—settled in Columbia county, in this State, and introduced the cotton gin, although Whitney claimed the credit of it and will probably always be known as the inventor of a machine which has produced such a marvelous revolution in the commerce of the world. Bull lived in Columbia county, and Whitney resided on the plantation of Gen. Green (of revolutionary memory) in Liberty county, two counties at that period considered very remote from each other and between them there was but little intercourse. Their inventions having the same object in view were nevertheless made without a knowledge of any preëxisting machine for ginning cotton. Bull used at first perpendicular saws, but very soon ascertained that circular saws were better adapted to his purposes and substituted them. Whitney obtained a patent for his invention and commenced suits against Bull and others who were using his gins in the United States Court, as the records of that Court at Savannah will, I presume, show. These suits were never tried, as it was understood that Bull was prepared to prove by reliable and incontrovertible testimony that his gin was his own invention and was no infringement of Whitney's rights under his patent; and I have Judge Bull's authority for saying that Whitney offered his father ten thousand dollars to suffer a judgment to be rendered against him, which he refused.

When Bull first put his gin in operation he ginned for the fourth, and excluded all male visitors, but females who were prompted by motives of curiosity to see it, were admitted. Some one who was a mechanic or a machinist introduced himself in the disguise of an old woman and with a walking stick on which its measure in inches was obscurely marked, obtained the dimensions of the machine and with the knowledge thus surreptitiously procured, constructed a gin on the same model.

That Whitney was entitled to the credit of the invention which he patented is probable, but that Bull was the first to introduce the saw gin—the prototype of the gin now in use—I have not the least doubt. Whitney's name has ever been and will always be connected with this great and important invention, and it is to be regretted that Bull's claim to the honor of an invention which has excited such a wonderful influence in controlling the commerce of the world, and has contributed so much to the comforts and the wants of mankind, cannot, owing to the lapse of years, be successfully vindicated.

After his extraordinary success in constructing a machine for ginning cotton, Bull went to New York and had two iron screws cast for pressing cotton. They were employed in the city of Augusta in repacking cotton for shipment. These were probably the first screws ever used baling cotton. Whatever doubt may exist in relation to Bull's claim to the invention of the gin, there is but little doubt but that he is entitled to the credit of the first packing screw.

Col. Dawson, of the Sulphur Springs in Meriwether county, remembers when Edward Lyon, who had been in Bull's service, built the first gin in Wilkes county. He thinks this occurred in the year 1806; and he remembers that Gilbert and Pruden had the first screws for packing cotton in that county, which were located in Washington, and made of cast iron. There are many Georgia who remember when the wooden screw was introduced. Previous to that time, nearly all the cotton made was packed in round bales without the agency of the screw. Such screws as were in use were made of cast iron.

Your correspondent does not seem to be aware of this fact, and I think it probable that the two he found on his plantation had been long since discarded and their place supplied with the safer and more economical wooden screw. He refers to their durability as a recommendation. It is true that no limit can be prescribed to the duration of cast iron, but in the shape of a screw it is, because of its brittleness, liable to break in exerting the immense power which is required of it in packing a bale of cotton, and when it does break it gives no premonition of the danger which menaces every one within reach of it. The age of cast iron screws has passed away, and I do not think that your correspondent, even with the aid of Mr. Finley, can revive it.

Pike County, Ga.; 1856. ANTIQUARY.

SECTION VII.—HISTORY OF THE COTTON GIN.

An esteemed correspondent, of Pike County, Ga., writes us as follows; and we commend his suggestions to the attention of Judge Andrews and others. The true history of this important invention, should be preserved:

Editors Southern Cultivator—Could you not induce some one who has a taste for home antiquities, to give you the history of the cotton gin? There are probably some living in Columbia county or in that part of the State whose recollection goes as far back as 1795. If so, they could shed much light on the subject. I should be pleased to know when Whitney's patent bears date, and whether the record of the United States Court at Savannah would not afford some information. Much of what is known on the subject is traditionary and unless collected will very soon be lost. There is an aged and an intelligent citizen of La Grange, by the name of Amos, who, Judge Bull informs me, worked in the shop of his father. From him and from other aged citizens much information could be collected. I know of no one who has more taste and capability for writing the history of the cotton gin than Judge Andrews, and I presume that at your suggestion he would favor the readers of the *Cultivator* with another communication on the subject.

Respectfully yours, W. D. A.

ELI WHITNEY,* THE INVENTOR OF THE COTTON GIN.

A STRUGGLE always presents a manly and inspiring spectacle. Man was made for action—and he cannot but sympathize with earnest and energetic action on the part of others. The struggle of brute force against brute force, is not without interest. The strife of mind with mind is nobler; nobler still is the struggle of mind with unwilling nature, when he is sternly resolved on wresting from her reluctant grasp the secret of her mystery. The interest increases just as the genius is commanding—as the obstacles are great and manifold—as the strife is protracted—and as the triumph is complete and final. If the struggle be for a worthy object, and that object be fully secured in some permanent benefit to mankind, which remains as its lasting memorial, it is nobler still.

For one or all of these reasons—the lives of "self-made men" have usually a peculiar charm. They are always read with an eager interest by the young and hopeful. Most of all are they favorite books with the young American. The structure of our government and society, gives leave to every man to make the most of himself. The buoyant and hopeful youth of our people, the boundless and undeveloped resources holding out so wide a field for effort, and the familiar spectacle of men, who, from the humblest origin, have risen by native energy to the highest stations of wealth and honor, these all combine to make the incidents of the life of such men the favorite reading of multitudes among us.

There are few lives of this class that present the elements

* Memoir of Eli Whitney, Esq. By Denison Olmstead, Professor of Natural Philosophy and Astronomy, Yale College. First published in the American Journal of Science, for 1832. New-Haven: Durne & Peck. 1846. pp. 80.

of higher, we might almost say, of more romantic interest, than the life of Eli Whitney. All the elements we have named are here present. There is great genius, adequately trained for its conflict. There is an object most noble and inspiring, and clearly contemplated by him as worthy his efforts—there is success the most complete and triumphant, in a result the value of which defies all computation; and there are obstacles enough to invigorate, to test, and develope the sternest heroism. We do not propose to give here an extended view of the character of Whitney, or a history of his life. Both of these have been ably done in the work of which we have given the title. There are, however, some facts and some personal traits, connected with the history of his greatest invention, which ought to be familiar to all our citizens.

Whitney was born in Massachusetts, at Westborough, in the year 1765. His father was a frugal, hard-working farmer, who had some taste for mechanics, as it would seem, having provided himself with a work-shop, which was stocked with a small supply of tools. This work-shop laid the foundations of Whitney's fame, and strengthened the decided genius for mechanics which he very early developed. From the earliest age at which he could handle tools, he was always in this shop. At about the age of twelve, he made a very tolerable violin, which was finished in all respects, and furnished very good music. This wonderful performance, for a boy of his age, and at that period in the history of our country, when the mechanic arts were so rude, in an interior country town too, as might reasonably be supposed, established his fame as a mechanic. From this time he was employed to repair violins, and to execute difficult jobs of various kinds, in all of which he seems to have been uniformly successful. At about this period, he took the opportunity, during the absence of his father at church, of prying into the mystery of his watch, which was to him a strange and unknown thing. Before he was aware of what he had done, he had taken it in pieces. But true to his genius, he attempted at once to put it together, and succeeded so perfectly, and so soon, that his father never suspected what he had done. At thirteen, he made a handsome table-knife, to supply the place of one of a well-finished set which had been broken, and succeeded so completely, that excepting the stamp upon the blade, for which he had not the necessary tools, it matched perfectly with the others. At the age of fifteen or sixteen, he proposed to his father, with characteristic enterprise, to commence the manufacture of nails, which were then made entirely by hand. It was in the midst of the war of the Revolution, and nails were scarce and dear. This enterprise was profitable, so profitable that after two years he determined to enlarge the business, and set off on a secret journey to find a suitable fellow-workman. After travelling forty miles he found his man, and returned—having called at every shop by the wayside, to gather from each all the information which he could in respect to the mechanic arts.

Such was Whitney in his boyhood; distinguished not only for his mechanical skill, joined with bold and self-relying enterprise, but also for a decided interest in the mathematics. His feelings were ardent, yet completely tempered and controlled by prudence.*

* When Mr. Whitney was eighteen years of age, he became distinctly conscious that he had not the control of his own mind—that his imagination was so fruitful and roving, and his temperament so excitable, that he could not command his attention. He at once set himself, by a deliberate effort, to gain the mastery of himself, and actually to hold his mind to a given point. The effort was trying—it cost him a whole night of struggle; but the victory was complete, and he felt ever after that his self-command was sufficient. He showed to his friends, all his life after, the results of this effort in the control of his attention, by which he could pass from one subject to another, be, as it were, entirely absorbed in it, and then take up the one which he had left, and find it just as he had left it.

From the age of nineteen, he formed the project of receiving a collegiate education, and though thwarted and delayed by influences at home, he adhered to this determination till four years after, at the age of twenty-three, when he was admitted to the Freshman class at Yale College. The expenses of his collegiate career were defrayed from his own industry, with temporary loans from his father. We regard this purpose, formed by such a young man, at so late a period of life, and carried through after so long a delay, as a decisive and striking indication of strong good sense, and a very elevated and comprehensive intellect. The quick and able mechanic is, of all men, the most likely to cherish an overweening sense of his own gifts, and to think that the peculiar skill in which he towers above the whole circle of his acquaintance, is the only knowledge worth possessing. The rewards and promises which hold out to such a man the allurement of speedy and brilliant success, are usually too exciting to be thrust forward into the dim future, for the sake of the unattractive studies of abstract science. What views Whitney entertained on this subject, or what particular consideration decided him upon a course so unusual, we are not informed. We record the fact as a decisive proof that his genius, though, from the first, daring and self-confident, was freighted with a large measure of foresight, comprehensiveness, and good sense. At college he was much interested in mathematical and philosophical studies, and constantly gave proof that his genius in invention and in practical mechanics was not in the least exhausted.

Thus ended the period of his preparation for the great work to which he was destined to apply his powers. This preparation was singularly complete. There was the earliest and brightest promise, answering completely to the word genius, as understood in its most peculiar and highest import—which genius had been rarely disciplined in those two opposite yet equally essential courses of training—the training of practical life and that of scientific studies. The descent of such a man upon the arena of great achievement, is as the appearance of a giant wearing a giant's panoply, either of which it is pleasant to look upon, but both of which united are splendid and imposing.

After leaving college, Mr. Whitney almost immediately went to the State of Georgia, for the purpose of fulfilling an engagement with a gentleman to reside in his family as a private teacher. On his way to Savannah, by ship, he had as a companion of his voyage, the widow of the then late Gen. Greene, so distinguished in the annals of our revolutionary history. On his arrival at Savannah, being but partially recovered from the small-pox, which he had by inoculation, he was invited by Mrs. Greene to spend a little time at her residdence at Mulberry Grove, near that city. He soon learned that another teacher had been employed in the place which he had expected. Mrs. Greene at once kindly and generously proposed to him to commence the study of the law under her hospitable roof, and to remain in her family as long as he should choose. He had not been long with her before he gave striking proofs of his mechanical ingenuity, which attracted the attention of Mrs. G., and led her to feel that Whitney could meet any exigency in which invention and skill of this kind were required. Not long after, Mrs. Green was visited by several gentlemen from Upper Georgia, principally officers who had served with her husband in the war. Of these were Majors Brewer, Forsythe, and Pendleton. They conversed largely upon the situation and prospects of agriculture in the opening upper country of the South, and expressed regret that no means had been devised to clear the upland cotton from the seed, saying that unless such a point could be attained, it was vain to raise cotton for the market. Mrs. Greene interrupted their conversation, by saying, "Gentlemen, apply to my young friend, Mr. Whitney, he can make anything." After showing them, as the results of his ingenuity, the various mechanical contrivances which he had devised and executed, she introduced him to the circle, who at once made known the object to be accomplished, and the difficulties which were in the way. Whitney, in reply, disclaimed any superiority of mechanical genius, and added, that he had never in his life seen either cotton or cotton seed. Mrs. Greene then said, "I have accomplished my aim. Mr. Whitney is a very deserving young man, and to bring him into notice was my object. The interest which our friends now feel for him, will, I hope, lead to his getting some employment to enable him to prosecute the study of the law." The interest of Mrs. Greene in this young and ingenious stranger, who had been fortuitously thrown in her way, deserves to be recorded in her honor. Such interest is not, we believe, uncommon, particularly at the hospitable home of the generous Southerner. It is rare that it meets with a reward so befitting, yet so splendid, as awaited Mrs. Greene, of having her name associated with the man and the invention which was destined to produce so striking a change on the interests and importance of the entire southern country.

Some of our northern readers may here, perhaps, need to be informed, that there are two kinds of cotton raised at the south—the one, the *Sea Island, the black seed* or *long staple* cotton; the other *the upland, green seed* or *short staple.* One of these species can be grown only upon the lowlands near the sea. Its fibre is long and fine; it can be separated from the seed with comparative ease, and it is used in the finer fabrics, as cambrics and muslins. This cotton was the only species that was extensively cultivated previous to Whitney's invention, and its growth was confined, as it is now, to rare and peculiar situations. Its price, per lb., is many times that of the other; and at this day, though a plantation fit for its culture is of rare value, yet the value of the entire production of this species is quite insignificant compared with that of the whole cotton crop of the Union.

The upland cotton can be raised on a large portion of the interior lands of the Southern States. Its fibre is short, and adheres tenaciously to the seed, and presented such difficulties in being cleaned, that the separation of a pound of cotton was esteemed the work of a day for a single hand; this circumstance alone interposing an insurmountable obstacle to its general or profitable culture. It may easily be understood, why the subject of a mechanical invention of this sort was esteemed so desirable by these gentlemen, residing, as they did, on the borders of this upper country—and being able to foresee truly, yet dimly, what immense results were hanging upon the possibility of such an invention.

The hint given to Whitney by these gentlemen was not lost upon him. The season for cotton in the seed was passed, but Whitney went to Savannah at once, and after a long search, at last lighted upon a small quantity; with this he returned to his temporary home, and communicated his intentions to Mr. Miller, who was then a teacher in the family, and afterwards married Mrs. Greene. A room was assigned to him, to which Mr. Miller and Mrs. Greene were the only persons who were admitted, or who knew anything of his project. His materials and tools were both limited; even the wire which he required could not be found at Savannah, and he was forced to draw it for himself. "Near the close of the winter, the machine was so nearly completed as to leave no doubt of its success." Mrs. Greene was naturally eager to communicate to her friends the fact of an invention which promised at once a crop suitable to the soil, occupation for their hands, and immense wealth, as the result of the extended culture of an article which had been thought of little worth. She invited to her house gentlemen of distinction from different parts of the State, and conducted her assembled guests to the room in which they saw with astonishment a machine which promised such splendid results for all their interests.

Mr. Whitney was at once urged to receive a patent. His reply was prophetic of what actually occurred, and was in substance, that the introduction of a new invention, the protection of it against encroachments of interested and unprincipled men, was an enterprise of doubtful success; and that rather than incur the hazards incident to it, he preferred to strive for the surer rewards attendant upon his contemplated profession. This remark was sagacious and prophetic, for after being overruled in his decision, and persuaded to embark his time and his energies to the introduction of his machine, under the protection of a patent, and after spending years of vexation and toil of body and spirit, in the effort to secure to himself some profitable return for his service to his country, he was forced to retire from the contest, with the persuasion that the effort had been a fruitless one; that what he gained by strife and determined perseverance, was no more than an equivalent for what he actually expended in the efforts to secure to hinself his rights; the actual loss of time in energy, which, if undistracted, might have been profitably directed to other pursuits—of health, and even of life, being reckoned only as a small item in the calculation.

His determination on this subject was changed principally, it is believed, by the agency of Mr. Miller, who entered into copartnership with him for the construction and vending of these machines, of which the profits were to be equally divided. The machine were to be patented. The necessary funds for the business was to be furnished by Miller. The instrument of this copartnership is dated May 27, 1793. But the invention could not be kept a secret till it should be protected by a patent. Great numbers of persons flocked from all parts of the State to see the new invention; and when it was not deemed prudent to allow access to the machine, the building was broken open by night and the new cotton-gin was carried off, so that before the model could be finished, and the letters patent could be secured, the invention was put in operation and several machines were constructed.

The petition for a patent was presented to Mr. Jefferson, the Secretary of State, June 20, 1793. Mr. Jefferson at once took a strong interest in the invention and its originator, and assured Mr. Whitney, that his request should be granted as soon as the model should be lodged at the patent office. In consequence of unavoidable delays, however, the patent was not secured in form till several months afterwards.

It would seem, that with the protection and power of a patent, for a machine so certain to be used and to increase the demand for itself by creating a new staple of the country, the road to affluence would be short and easy. But events issued far otherwise. Obstacles the most trying and depressing at once arose, and with them the persevering spirit of Whitney was summoned to contend during the entire period of fourteen years allowed him, by the patent law. The history of this period is a history of vexation, disappointment, and of wrong—not indeed of wrong on the part of those who occasioned it; that was in all or in most cases, deliberate and known, but which to him who suffered it, brought all the painful consequences of such wrong.

We cannot go into a detailed history of the obstacles against which he was called to struggle. A few of the facts only can be given, and these only in a summary way. A prominent cause of these may be found in the plan adopted by the partners, to control and manage the business themselves, by erecting machines in all parts of the country, and to *gin* the cotton at a certain rate per pound, or to buy the cotton before it was separated, and then to control, to a certain extent, the whole crop in the market. It is easy to see how tempting this plan must have seemed to the eyes of Whitney's sanguine partner; how certain to realize immense and sudden wealth; and how likely, on the other hand, to arouse the jealous antagonism of the planters on whose eyes were also beginning to dawn bright visions of the wealth, which they too might realize through this new channel. A proposition which might seem to appropriate too large a portion of the profits would be likely to sharpen their doubts of the originality of the invention, and to blunt their sense of justice in using the cotton gin wherever it could be found, or whoever might be in law its proprietor. The recent enactments of the patent law, for it was passed in the early part of 1793, the year in which Whitney's application was made, might also have contributed to the difficuties with which Whitney was called to contend. The law which secures to an inventor his rights is less readily appreciated and respected, than one which guards national property, or bodily life. Besides, under a government so recent as that of the Union, the impost of such a law was yet to be settled by actual decisions, and its applications to be tested by verdicts of juries. At this time also money was scarce, and rates of interest were high, so that at the time when a loud call was beginning to be made for the machines, which were to be manufactured by Miller and Whitney, the partners were themselves embarrassed for the want of capital with which to make them. Then again, at one time, Whitney was prostrated by disease also; and at a second, both being important crises in the fortunes of the infant enterprise, his workmen also were disabled by a fatal epidemic, in July 1794, just at the time when the first cotton crop was maturing for the new machine, and when Miller was writing to Whitney that within a year from that time from fifty to a hundred gins must be completed and transported to the south. He adds, "The people of the country are running mad for them, and much can be said to justify their importunity. When the present crop is harvested, there will be a real property of at least fifty thousand dollars lying useless, unless we can enable the holders to bring it to market."

Early in the year following, on arriving at New Haven, from New York, where he had been detained by a lingering illness, he was informed that on the day before, his shop, with all his machinery and papers, had been consumed by fire. At this time, under the pressure of necessity, (which is the mother of all sorts of inventions, some of which are not the most honest) the necessity arising from the increased culture of cotton, which had been immensely stimulated by the prospect from the invention, two rival machines appeared in the field to dispute the claims of Whitney's. The one of these was the *roller-gin*, which, though it executed the work of cleansing the cotton very imperfectly, yet, in the exigency, found many advocates. At all events, it diverted the attention of the public from Whitney's patented machine, and weakened their moral sense in respect to any peculiar claims on his part. The other was the *saw-gin*, which applied one of the principles peculiar to Whitney's, with this difference, that the teeth were cut from a continuous plate of metal, instead of being inserted as wires. This idea, by the way, had occurred early to Whitney, as was established afterwards by legal proof. Here, then, was a machine which was really his, and against which he brought his suits, and at last enforced the rights of his patent. But as yet, and for years, while the question was undecided, this was as good as the patent one, and many an honest man might think himself justified in using it. This machine made its appearance in 1795. The year after—for each year brought to Whitney its new calamity—the fatal intelligence was brought from England that the manufacturers rejected the cotton cleaned by the gin, because the staple was supposed to be injured. This at once lowered the price of his cotton in the market, and gave boldness to the trespassers upon his rights. At this moment the company had thirty gins stationed at eight different places in the State of Georgia, and $10,000 invested in real estate, for the purposes of their enterprise. Near the close of the following year the stout heart of Whitney begins to yield, and he writes as follows: "I have labored hard against the strong current of disappointment, which has been threatening to carry us down the cataract; but I have labored with a shattered oar, and struggled in vain, unless some speedy relief is obtained." And in the same letter—"I have sacrificed to it [our business] other objects, from which, before this time, I might certainly have gained twenty or thirty thousand dollars." During this year, however, the reports from England, in respect to the quality of the cotton from the new machines were entirely reversed, and a preference began to be given to it over every other in the market.

But the peculiar calamity of this year (1797) was, that the first trial of their patent at law which could be obtained, issued against them. Notwithstanding the charge of the judge was pointedly in their favor, and the defendants expected a verdict of heavy damages, the plaintiffs lost their case, and an application for a new trial was denied. Thus, after four years had been consumed in a protracted effort to test the validity of the patent, at the first issue that was joined there was an entire failure. Vigorous efforts were made to bring another suit to trial at Savannah, in the year following, 1798. Witnesses were assembled from various and distant parts of the State, and at great expense; but the judge did not appear, and the trial was deferred. About a year after, Mr. Miller seems to have given up all hopes of defending the patent in the State of Georgia; and it may easily be imagined how, in a new State, under the new federal government, with this delay of an enforcement of the right for years, the right might be worthless, especially as the payment of claims which were allowed might be put off for four years, when they would expire by the then existing statute of limitations. In South Carolina a different plan was adopted, at the suggestion of influential planters. The proprietors of the cotton gin proposed to the legislature, to relinquish their patent-right for the use of the citizens of the State, in consideration of $100,000. This proposal was made in December, 1801. The legislature offered $50,000; $20,000 to be paid in hand, and the remainder in three annual instalments of $10,0C0 each. This offer was accepted, though Mr. Whitney writes in respect to it: "This is selling the right at a great sacrifice. If a regular course of law had been pursued, from two to three hundred thousand dollars would undoubtedly have been recovered. The use of the machine here is amazingly extensive, and the value of it is beyond calculation. It may, without exaggeration, be said to have raised the value of seven eighths of all the three Southern States from fifty to a hundred per cent. We get but a song for it in comparison with the worth of the thing; but it is *securing* something."

Thus, after more than seven years after the patent was issued, and when its time to run had more than half expired, it returned to its owners twenty thousand dollars in hand, with the promise of thirty thousand more. It would be an insult to the common sense of our readers, to suppose it necessary to argue in detail the proposition that far more than this might easily have been sunk, before this return, in the expenses of unsuccessful suits, and in the unproductive capital that had been invested in the enterprise of manufacturing and working the gins, which, having no protection of law, would, of course, be of little worth to their owners. A year after the bargain with the State of South Carolina, in December, 1802, the right was sold to North Carolina, the legislature imposing an annual tax of two shillings and sixpence on every *saw*, for the benefit of the patentee. Some of the gins contained forty *saws*, and the tax was to be collected for five years. The cultivation of cotton was at that time limited; but, in consideration of the use that was made of the gin, this was thought to be the most liberal compensation that was offered from any service. Another sale, on similar terms, was made to the State of Tennessee the year following, 1803, the legislature imposing an annual tax of 37½ cents on every saw, for four years.*

This bright dawning of a better day, though deferred so long, was not unclouded, even when it at last appeared. While Mr. Whitney was negotiating with North Carolina, he learned that South Carolina had repented of its just resolve—had suspended the payment of the balance due him, and had instituted a suit for the recovery of what he had already received. At about the same time, the Governor of Georgia, in his annual message, took very decided ground against any grant to the

* We have before us the *Aurora and General Advertiser*, published daily at Frankford, dated Sept. 3, 1802, in which there is a detailed account of a meeting of the citizens of Nashville, Tennessee, July 21, of which General Andrew Jackson was chairman, and the account of which is signed by him as such. After a preamble, the meeting resolved, "That it will tend much to the agricultural and commercial interests of this State, that the legislature, at their next session, purchase the patent-right of the said saw gin, for the use and benefit of its citizens, and lay a tax on the makers and users of said gins, to discharge the said sum which may be contracted, to be given to the patentees for the patent-right aforesaid," &c., &c.

patentee. A committee of the legislature, to whom this part of the message was referred, reported strongly in its favor, and urged united action, on the part of the then Southern States, [Georgia, South and North Carolina, and Tennessee,] to resist the grievance of the patent, as likely to depress "the culture and cleaning of the precious and increasing staple," or to petition the General Government to make just compensation to the inventor. Tennessee followed the example of South Carolina, and suspended the payment of the stipulated tax; while North Carolina adopted a resolution "that the contract ought to be fulfilled with punctuality and good faith." The grounds of this hostile movement on the part of South Carolina, were, first, a technical failure on the part of the patentees to fulfil some stipulation in the contract; and, second, a real suspicion of the originality of the invention, it being contended all over the South that such a machine had been seen in use in Switzerland forty years before, for the purpose of picking rags to make lint and paper. The tide in South Carolina, however, soon turned, and in the year 1804 the legislature confirmed their original contract. About this time Mr. Miller died, leaving Whitney to struggle alone. He at last triumphed in the State of Georgia, and obtained a decision vindicating his patent, in December, 1807, just about a year before the expiration of his right; and the year following, as the right was expiring, two other suits were gained. In all of them the originality of the invention was triumphantly established, and the rights of the patentee were clearly asserted. These decisions were, however, too late. But the fact that they had been delayed so long, was not the most painful aggravation. This arose from the circumstance that, for nearly thirteen years, his time and physical strength, and his mental energy, were absorbed in this vexatious enterprise of contending for his rights. More than *sixty* suits had been brought, before a single decision was obtained which vindicated these rights. He had made six journeys from New-Haven to Georgia—several by land—some of which involved the severest exposure to his health and the hazard of his life. It is the testimony of a gentleman who was intimately acquainted with Mr. Whitney's affairs at the South, and whom he consulted as a legal adviser, that, "in all his experience in the thorny profession of the law, he had never seen such a case of perseverance under such persecutions; nor," he adds, "do I believe that I ever knew any other man who could have met them with equal coolness and firmness, or who would finally have obtained even the partial success which he had. He always called on me in New York, on his way South, when going to attend his endless trials, and to meet the mischievous contrivances of men who seem inexhaustible in their resources of evil. Even now, after thirty years, my head aches to recollect his narrations of new trials, fresh disappointments, and accumulated wrongs."

In 1812, Mr. Whitney applied to Congress for the renewal of his patent. His memorial is long and able, stating the painful story of his struggles, and the little profit that he had received. He says, that "from no State had he received the amount of half a cent per pound on the cotton cleaned with his machine in one year. Estimating the value of the labor of one man at twenty cents per day, the whole amount which had been received by him for his invention, was not equal to the value of the labor saved in *one hour* by his machines then in use in the United States." "The invention has already trebled the value of the land through a great extent of territory, and the degree to which the cultivation of cotton may still be augmented, is altogether incalculable." "In short," to quote the language of Judge Johnson, of South Carolina, "if we should assert that the benefits of this invention exceed one hundred millions of dollars, we can prove the assertion by correct calculation." "There is no probability that the patentee, if the term of his patent were extended for twenty years, would ever obtain for his invention one-half as much as many an individual will gain by the use of it. Up to the present time, the whole amount of what he has acquired from this source, (after deducting his expenses,) does not exceed one-half the sum which a single individual has gained by the use of the machine in one year. It is true that considerable sums have been obtained from some of the States where the machine is used; but no small portion of these sums has been expended in prosecuting his claim in a State where nothing has been obtained, and where his machine has been used to the greatest advantage."

This memorial was sustained by several distinguished individuals from the cotton-growing districts, but it was not granted, and it was never renewed.

This closes the history of the cotton gin, so far as the connection of its inventor with it is concerned. But it does not finish the history of Whitney's services to his country. In the year 1798, despairing of anything of consequence from his cotton gin, Whitney embarked in a new enterprise, that of manufacturing arms for the government. He was ignorant of the details of the business, and as yet had no works at command, no capital, and no workmen, and yet he ventured with clear and well sustained confidence in his own resources, into a business that was complicated, embarrassing, and new. The result was, the most complete success—or rather it might be said that he created this branch of manufactures anew. His methods were entirely new and peculiar, both in the allotment of the work, and in doing very much by machinery, of various and complicated construction, that had hitherto been done by the file, with an experienced eye and hand. The result was, that the several pieces of a musket, made at his establishment, were so exactly alike, that the smallest screw or spring fitted for one, is equally fitted for any and every other. These improvements were introduced against much skepticism and many obstacles, into all the public and private armories of the Union; and it is mainly owing to Whitney that the manufacture of arms by this government, is unsurpassed in any public armories in the world. There is, perhaps, no exhibition in the manufacture of metals, that is more beautiful and exciting, than that furnished in the armories at Springfield and Harper's Ferry, and it is Eli Whitney to whom these improvements are owing. It was admitted by a former Secretary of War, in conversation with Mr. Whitney, that the annual saving, years ago, at the public armories, in consequence of Whitney's improvements, was more than $25,000 a year. It was from this business that Whitney derived the greater portion of the estate which he accumulated; but it is believed that this estate amounted to hardly more than he enabled his country to save, at this moment, in her armories, in a single year.

These improvements are not necessarily confined to the manufacture of arms, but are applicable, and have been applied to the working of metals of every kind, and Whitney's memorials may be said to exist in every machine-shop in the land.

The incidental benefits conferred by him upon the entire circle of the mechanic arts, and the interests connected with them, are not lightly to be esteemed. The elevation of these arts, and of those connected with them, in the estimation of the public, by the devotion to their advancement of so high a genius, and by the manifest demonstration of the fact, that the highest and most thorough mastery of science strengthens rather than weakens the hand of art, are felt, after the man who has exerted these influences is gone, and form no inconsiderable item in the sum total of the good which he has conferred upon his race. It was certainly true in the case of Whitney, that his personal influence, and the weight of his example, were thus felt in his life-time, and have not ceased since his death. It is owing to his influence, and that of Hillhouse, men of singular kindred spirits, in respect to the devotion of energy and genius to the general improvement of the outward interests of man, that the City of Elms is so attractive to the eye of the stranger, in the outward indications of neatness and taste, which pertain to elevate every dwelling, and are seen along every street, and that these external appearances are found, on a close observation, to be so sure an index of the intelligent industry, and the frugal thrift of the citizens.

But the great gift of Whitney to his country and his race, was the gift of the cotton gin; and it is for this that he will deserve to be cherished longest in their honorable remembrance, and in their grateful homage. We have seen that there was something singularly interesting in the manner and circumstances of the invention. By far the greater number of inventions, come of accidental suggestion, or are completed by gradual improvements, or result from the application of machinery already in being to a new purpose; but in the case of this of Whitney's, there was the contemplation of a great desideratum, proposed as the worthy subject for a trained and powerful genius; then there was the cheerful devotion of the mind to meet this want, and then the speedy, the easy, and the successful triumph.

But the value of the gift deserves our consideration. We have spoken of it as a gift of Whitney to his country and to his race. What then did Whitney give to his country?

He gave to his country, directly, all the increased value which the public lands of the cotton-growing States have received by the invention. So soon as this invention was made known, and its adequacy was fully established, the inland districts of the south and south-west at once rose immensely in value, and the extensive public lands of the United States rose with them. Whatever, therefore, the government, as a direct owner of property, which it offers for sale, has received, or is yet to receive from this advance upon its property, that has been and will be put into her treasury, by the gift of Eli Whitney; of what the value of his gift to her in this form has been, and is to be, some idea may be formed from a remark of Mr. Jefferson to Mr. Whitney, not long after the purchase of Louisiana, that the increased value of the lands of the United States, in consequence of the cotton gin, had at that time (this was ten years after the invention), been more than sufficient to pay the cost of that territory. How many times fifteen millions of dollars have since been added to the value of the public domain, by the increased culture of cotton, and the widening market for it, can neither be estimated nor conjectured.

Whitney gave to his country its greatest staple production, and the means of an extensive and profitable trade with England. Though the cotton plant had been known before the days of Herodotus, and though the green seed or upland cotton had been known from this early period, yet, as an article of commerce, it never had been known till this method of cleaning was discovered. A few statistics need only be given to show the immense value of the production, which was created by this invention, and of the trade which has grown out of it. In the year 1791, the whole cotton crop of the United States was 2,000,000 lbs. In 1845, it was more than 1,000,000,000. In 1791 the United States produced $\frac{1}{245}$ of the cotton produced in the world. In 1845, it produced more than $\frac{7}{8}$ of the product of the world.

1793, the year of the invention, the whole crop was

5,000,000 lbs., and the quantity exported was 487,600 lbs. In 1794, the year *after*, the crop was 8,000,000, and the exportation was 1,601,760 lbs. In 1800, six years after, the crop was 35,000,000, and 17,789,803 were exported. In 1810, two years before Whitney applied for the renewal of his patent, the crop was 85,000,000 lbs.; and of *upland* cotton, 84,657,384 lbs. were exported. In 1845, the entire crop was 1,029,850,000 lbs.; 862,580,000 lbs. were exported, and 167,270,000 lbs. were consumed at home. Cotton has been for many years not only one of the staples but *the* great staple for export. For many years past, it has constituted from one half to seven-tenths of the entire exports of the Union.

These facts speak for themselves. They tell us that the planters of the south owe it to the cotton gin, that, for half a century past, they have been able to raise and send to market their great staple, and that it is to Whitney that they are indebted for the great estates they have accumulated, and the ample incomes which they have so generally expended. Whatever wealth the country has received in the increase of individual estates, the country owes to the inventor of the cotton gin. These statistics tell us, that whatever has been made by the immense trade between this country and its great customer on the other side of the sea, is mainly owing to the same invention. The shipper of the cotton owes to him the profits on his freights. The importer the profits on goods, which he has been able to buy with cotton; and the government, the revenue which she has exacted on these goods; as well as the immense advantage which she has gained from having so abundant a staple of her own, with which to pay for the imports which she has received. The manufacturers of cotton too, whether at the north, they drive their magnificent establishments, receive their splendid dividends; or whether at the south, they are inspired by the hope of the same success, in their vigorous, though infant enterprises; the busy troops of operators whom they pay, and the neighboring farmers who find a ready market for their produce, all owe it to the same genius, that they have the *material* art, of which to bring these larger or humbler returns. All that prosperity, too, which results from the combined and harmonious working of the producing, the commercial, and the manufacturing interests, as far as these interests have been dependent on the cultivation of cotton, has received its impulse from this invention, and owes its acknowledgment to Eli Whitney.

We have seen what Whitney has given to his country. The question is very natural, what has he received *from* that country? It was his own testimony, as asserted by an intimate friend, Professor Silliman, and that testimony was given in his hearing near the close of life, that the disease, which cut short his life, was brought on by exposure and fatigue during the last of his land journeys, to assert his just claims so long injuriously frustrated. He received then, first of all, the termination of his life in the midst of those domestic enjoyments which had been so long deferred in consequence of the delay of his just rights, and in the possession of that fortune, the acquisition of which he had been forced to put off till the noon of his days.

So far as a pecuniary return is concerned, he received *nothing;* for it was also his dying testimony, that "all he had received for the invention of the cotton gin had not more than compensated him for the enormous expenses which he had incurred, and for the time which he had devoted, during many of the best years of his life, in the prosecution of this subject." On such a subject as this, Mr. Whitney was not likely to exaggerate; his mind was too self-possessed, and his integrity too uncorrupt to allow him to yield to the gloom of disappointment, or the violence of passion. His cheerful application to new fields of enterprise; his ready and generous forwardness to serve his friends, his country, and his race, with no prospect of return, and the courteous hospitality with which he received and returned the warm esteem of gentlemen from all parts of the Union, as well as the fact that this esteem was so lavishly bestowed upon him—all show that his views of this subject were neither morbid nor selfish. When, therefore, he said, as he did deliberately, that he "felt that his just claims on the cotton-growing States, especially on those that had made him no returns for this invention, so important to this country, were still unsatisfied, and that both justice and honor required that compensation should be made"—we should feel assured that his testimony but expressed the truth in the case—if all the particulars which we have enumerated did not both suggest and confirm the same conclusion.

It is not necessary to go at lenghth into the reasons of his failure to receive the just compensation for his eminent services. Many can be imagined in the then infant state of the country, and the unsettled judgments of men in regard to the rights of discoverers, and the unequal action of patents, and in their jealous opposition to monopolies, without supposing a decided and deliberate purpose to defraud or wrong a man from whom the gotton-growers had received their all. We are quite certain that the State of Georgia, at this moment, would be as far from such injustice as any other in the Union. And if the question were presented to her now, whether she owed no debt to the inventor, she could not, in the view of her whitening cotton fields, and in the hearing of the noise of *her own* cotton mills, but generously acknowledge the obligation. It is not, however, an obligation for any one State. The whole Union is too much indebted to the great invention to be content to leave the obligation to be cancelled by any one of its sisterhood. The name of Whitney is too intimately associated with her honor, and with her unexampled growth and prosperity, to be remembered by her with any other than the profoundest gratitude.

We do not approve of lavish or indiscriminate testimonials to the honor of the living or the deceased, who have deserved well of their country; but that such a testimonial ought to be rendered to such a man, who has added uncounted millions to her wealth, is too clear to be argued. We are bold to say, that to no man, whether living or dead, does she owe more for her physical prosperity and wealth, than to the subject of these remarks. We trust the time may come when an opportunity will be furnished to repay this obligation, and the name of Whitney shall not be coupled with the ingratitude or neglect of this great and free people.

Upon his tomb-stone there is the following appropriate inscription:—"ELI WHITNEY, the inventor of the cotton gin. Of useful science and arts, the efficient patron and improver. In the social relations of life, a model of excellence. While private affection weeps at his tomb, his country honors his memory." *His country honors his memory!* Let it be seen that she does, not by idly bending over his tomb, nor by lauding his name by verbal adulation, but with generous and united zeal testifying to his family, some substantial token, that she appreciates the genius and services of the man who has contributed so much to her prosperity.—*Democratic Review.*

All the Books on this Catalogue sent by mail, to any part of the Union, free of postage, upon receipt of price.

BOOKS FOR THE COUNTRY,

PUBLISHED BY

C. M. SAXTON & COMPANY,

140 FULTON STREET, NEW YORK,

SUITABLE FOR

SCHOOL, TOWN, AGRICULTURAL, AND PRIVATE LIBRARIES.

Downing's (A. J.) Landscape Gardening; . . . $3 50

A Treatise on the Theory and Practice of Landscape Gardening. Adapted to North America, with a view to the improvement of Country Residences, comprising Historical Notices and General Principles of the Art. Directions for Laying out Grounds and Arranging Plantations, the Description and Cultivation of Hardy Trees, Decoration Accompaniments to the House and Ground, the Formation of Pieces of Artificial Waters, Flower Gardens, etc., with Remarks on Rural Architecture. Elegantly illustrated, with a Portrait of the Author. By A. J. Downing.

Downing's (A. J.) Rural Essays 3 00

On Horticulture, Landscape Gardening, Rural Architecture, Trees, Agriculture, Fruit, with his Letters from England. Edited, with a Memoir of the Author, by George Wm. Curtis, and a letter to his friends by Frederika Bremer; and an elegant steel Portrait of the Author.

Practical Fruit, Flower, & Kitchen Gardener's Companion, 1 00

With a Calendar. By Patrick Neill, LL.D., F.R.S.E., Secretary of the Royal Caledonian Horticultural Society. Adapted to the United States, from the fourth edition, revised and improved by the author. Edited by G. Emerson, M. D., Editor of "Johnson's Farmer's Encyclopedia." With Notes and Additions by R. G. Pardee, author of "Manual of the Strawberry Culture." With illustrations.

Munn's (B.) Practical Land Drainer; 50

Being a Treatise on Draining Land, in which the most approved systems of Drainage are explained, and their differences and comparative merits discussed; with full Directions for the Cutting and Making of Drains, with Remarks upon the various Materials of which they may be composed. With many illustrations. By B. Munn, Landscape Gardener.

Elliott's Am. Fruit-Grower's Guide in Orchard and Garden 1 25

Being a Compend of the History, Modes of Propagation, Culture, &c., of Fruit, Trees and Shrubs, with descriptions of nearly all the varieties of Fruits cultivated in this country; and Notes of their adaptation to localities, soils, and a complete list of Fruits worthy of cultivation. By F. R. Elliott, Pomologist.

Pardee (R. G.) on Strawberry Culture 60

A Complete Manual for the Cultivation of the Strawberry; with a description of the best varieties.

Also, Notices of the Raspberry, Blackberry, Currant, Gooseberry and Grape; with directions for their cultivation, and the selection of the best varieties. "Every process here recommended has been proved, the plans of others tried, and the result is here given." With a valuable Appendix, containing the observations and experience of some of the most successful cultivators of these fruits in our country.

Dana's Muck Manual for the use of Farmers . . $1 00

A TREATISE ON THE PHYSICAL AND CHEMICAL PROPERTIES OF SOILS, the Chemistry of Manures; including also the subject of composts, artificial manures and irrigation. A new edition, with a chapter on Bones and Superphosphates.

The Stable Book 1 00

A TREATISE ON THE MANAGEMENT OF HORSES, in relation to Stabling, Grooming, Feeding, Watering and Working, Construction of Stables, Ventilation, Appendages of Stables, Management of the Feet, and Management of Diseased and Defective Horses. By JOHN STEWART, Veterinary Surgeon. With Notes and additions adapting it to American Food and Climate. By A. B. ALLEN, editor of the American Agriculturist.

Chorlton's Grape-Grower's Guide 60

INTENDED ESPECIALLY FOR THE AMERICAN CLIMATE. Being a Practical Treatise on the cultivation of the Grape Vine in each department of Hot House, Cold Grapery, Retarding House, and Out-door Culture. With plans for the construction of the requisite buildings, and giving the best methods for heating the same. Every department being fully illustrated. By WILLIAM CHORLTON.

White's (W. N.) Gardening for the South; . . . 1 25

OR, THE KITCHEN AND FRUIT GARDEN, with the best methods for their Cultivation: together with hints upon Landscape and Flower Gardening; containing modes of culture and descriptions of the species and varieties of the Culinary Vegetables, Fruit Trees and Fruits, and a select list of Ornamental Trees and Plants, found by trial adapted to the States of the Union south of Pennsylvania, with Gardening Calendars for the same. By WM. N. White, of Athens, Georgia.

Eastwood (B.) on the Cultivation of the Cranberry . . 50

WITH A DESCRIPTION OF THE BEST VARIETIES. By B. EASTWOOD, "Septimus" of the New York Tribune.

Johnson's (Geo. W.) Dictionary of Modern Gardening . 1 50

WITH ONE HUNDRED AND EIGHTY WOOD CUTS. Edited, with numerous additions, by DAVID LANDRETH, of Philadelphia.

Johnston's (J. F. W.) Catechism of Ag. Chemistry & Geol. 25

By JAMES F. W. JOHNSTON, M.A., F.R.SS.L. and E., Honorary Member of the Royal Agricultural Society of England, and author of "Lectures on Agricultural Chemistry and Geology." With an Introduction by JOHN PITKIN NORTON, M.A., late professor of Scientific Agriculture in Yale College. With Notes and Additions by the Author, prepared expressly for this edition, and an Appendix compiled by the Superintendent of Education in Nova Scotia. Adapted to the use of schools.

Johnston's (J. F. W.) Elements of Ag'l Chemistry . 1 00

AND GEOLOGY. WITH A COMPLETE ANALYTICAL AND ALPHABETBETICAL Index, and an American Preface. By Hon. SIMON BROWN, Editor of the "New England Farmer."

Johnston's (James F. W.) Agricultural Chemistry . 1 25

LECTURES ON THE APPLICATION OF CHEMISTRY AND GEOLOGY TO Agriculture. New edition, with an Appendix, containing the Author's Experiments in Practical Agriculture.

Smith's (C. H. J.) Landscape Gardening, Pleasure Grounds 1 25

WITH PRACTICAL NOTES ON COUNTRY RESIDENCES, VILLAS, PUBLIC

Parks and Gardens. By CHARLES H. J. SMITH, Landscape Gardener and Garden Architect, &c. With Notes and Additions by LEWIS F. ALLEN, author of "Rural Architecture," &c.

The author, while engaged in his profession for the last eighteen years, has often been requested to recommend a book which might enable persons to acquire some general knowledge of the principles of Landscape Gardening.

The object of the present work is to preserve a plain and direct method of statement, to be intelligible to all who have had an ordinary education, and to give directions which, it is hoped, will be found to be practical by those who have an adequate knowledge of country affairs.

Norton's (John P.) Elements of Scientific Agriculture; $0 60

OR, THE CONNEXION BETWEEN SCIENCE AND THE ART OF PRACTICAL Farming. (Prize Essay of the New York State Agricultural Society.) By JOHN P. NORTON, M. A., Professor of Scientific Agriculture in Yale College. Adapted to the use of Schools.

Nash's (J. A.) Progressive Farmer 60

A SCIENTIFIC TREATISE ON AGRICULTURAL CHEMISTRY, THE GEOLOGY of Agriculture, on Plants and Animals, Manures and Soils applied to Practical Agriculture; with a Catechism of Scientific and Practical Agriculture. By J. A. NASH.

Allen (J. Fisk) on the Culture of the Grape 1 00

A PRACTICAL TREATISE ON THE CULTURE AND TREATMENT OF THE Grape Vine, embracing its history, with directions for its treatment in the United States of America, in the open air and under glass structures, with and without artificial heat. By J. FISK ALLEN.

Mysteries of Bee-keeping Explained; 1 00

BEING A COMPLETE ANALYSIS OF THE WHOLE SUBJECT, consisting of the Natural History of Bees; Directions for Obtaining the greatest amount of Pure Surplus Honey with the least possible expense; Remedies for losses given, and the Science of Luck, fully illustrated; the result of more than twenty years' experience in extensive Apiaries. By M. QUINBY.

American Bee-keeper's Manual; 1 00

BEING A PRACTICAL TREATISE ON THE HISTORY AND DOMESTIC ECONOmy of the Honey Bee, embracing a full illustration of the whole subject, with the most approved methods of managing this Insect, through every branch of its Culture; the result of many years' experience. Illustrated with many engravings. By T. B. MINER.

The Cottage and Farm Bee-keeper; 50

A PRACTICAL WORK, by a Country Curate.

Weeks (John M.) on Bees.—A Manual; 50

OR, AN EASY METHOD OF MANAGING BEES IN THE MOST PROFITABLE manner to their owner; with infallible rules to prevent their destruction by the Moth. With an Appendix by WOOSTER A. FLANDERS.

The Rose; 50

BEING A PRACTICAL TREATISE ON THE PROPAGATION, CULTIVATION, and management of the Rose in all Seasons; with a list of Choice and Approved Varieties adapted to the Climate of the United States; to which is added full directions for the Treatment of the Dahlia. Illustrated by engravings.

Buist's (Robert) American Flower-Garden Directory; . 1 25

CONTAINING PRACTICAL DIRECTIONS FOR THE CULTURE OF PLANTS in the Flower-Garden, Hot-House, Green-House, Rooms, or Parlor Windows, for every Month in the Year; with a description of the Plants most desirable in each, the Nature of the Soil and Situation best adapted to their Growth, the Proper Season for Transplanting, &c.; with Instructions for erecting a Hot-House, Green-House, and

Laying out a Flower-Garden; the whole adapted to either large or small Gardens; with Instructions for Preparing the Soil, Propagating, Planting, Pruning, Training, and Fruiting the Grape Vine.

Buist's (Robert) Family Kitchen Gardener; $0 75

CONTAINING PLAIN AND ACCURATE DESCRIPTIONS OF ALL THE DIFferent Species and Varieties of Culinary Vegetables, with their Botanical, English, French, and German names, alphabetically arranged, and the best mode of cultivating them in the garden or under glass; also Descriptions and Character of the most Select Fruits, their Management, Propagation, &c. By ROBERT BUIST, author of the "American Flower-Garden Directory," &c. Paper 50 cents.

The American Florist's Guide; 75

COMPRISING THE AMERICAN ROSE CULTURIST AND EVERY LADY HER own Flower Gardener.

Every Lady Her own Flower Gardener; 50

ADDRESSED TO THE INDUSTRIOUS AND ECONOMICAL ONLY; containing Simple and Practical Directions for Cultivating Plants and Flowers; also, Hints for the Management of Flowers in Rooms, with brief Botanical Descriptions of Plants and Flowers. The whole in plain and simple language. By LOUISA JOHNSON.

The American Agriculturist; 10 00

BEING A COLLECTION OF ORIGINAL ARTICLES ON THE VARIOUS Subjects connected with the Farm, in ten vols. 8vo., containing nearly four thousand pages.

The Complete Farmer and American Gardener; . . 1 25

RURAL ECONOMIST, AND NEW AMERICAN GARDENER, containing a Compendious Epitome of the most Important Branches of Agricultural and Rural Economy; with Practical Directions on the Cultivation of Fruits and Vegetables; including Landscape and Ornamental Gardening. By T. G. FESSENDEN. 2 vols. in one.

Fessenden's (T. G.) American Kitchen Gardener; . . 50

CONTAINING DIRECTIONS FOR THE CULTIVATION OF VEGETABLES and Garden Fruits.

Dadd's (Geo. H.) American Cattle Doctor; 1 00

CONTAINING THE NECESSARY INFORMATION FOR PRESERVING THE health and curing the diseases of Oxen, Cows, Sheep, and Swine, with a great variety of original receipts, and valuable information in reference to Farm and Dairy management, whereby every man can be his own Cattle Doctor. The principles taught in this work are, that all medication shall be subservient to nature—that all medicines must be sanative in their operation, and administered with a view of aiding the vital powers, instead of depressing, as heretofore, with the lancet or by poison. By G. H. DADD, M.D., Veterinary Practitioner.

Browne's (D. Jay) Field Book of Manures: . . . 1 25

OR, AMERICAN MUCK BOOK; Treating of the Nature, Properties, Sources, History, and Operations of all the Principal Fertilizers and Manures in Common Use, with Specific Directions for their Preservation, and Application to the Soil and to Crops; drawn from Authentic Sources, Actual Experience, and Personal Observation, as combined with the leading Principles of Practical and Scientific Agriculture. By D. JAY BROWNE

Randall's (H. S.) Sheep Husbandry; 1 25

WITH AN ACCOUNT OF THE DIFFERENT BREEDS, and general directions in regard to Summer and Winter Management, breeding, and the treatment of diseases, with portraits and other engravings. By HENRY S. RANDALL.

Blake's (Rev. John L.) Farmer at Home. 1 25

A FAMILY TEXT BOOK FOR THE COUNTRY; being a Cyclopedia of

Agricultural Implements and Productions, and of the more Important Topics in Domestic Economy, Science and Literature; adapted to Rural Life. By Rev. John L. Blake, D. D.

Youatt and Martin on Cattle; $1 25

Being a Treatise on their Breeds, Management and Diseases, comprising a full History of the Various Races; their Origin, Breeding, and Merits; their capacity for Beef and Milk. By W. Youatt and W. C. L. Martin. The whole forming a complete Guide for the Farmer, the Amateur, and the Veterinary Surgeon, with 100 illustrations. Edited by Ambrose Stevens.

Youatt on the Horse. 1 25

Youatt on the Structure and Diseases of the Horse, with their Remedies. Also, Practical Rules for Buyers, Breeders, Smiths, &c. Edited by W. C. Spooner, M.R.C.V.S. With an account of the Breeds in the United States, by Henry S. Randall.

Youatt and Martin on the Hog. 75

A Treatise on the Breeds, Management and Medical Treatment of Swine, with directions for Salting Pork and Curing Bacon and Hams. By Wm. Youatt, R.S. and W C. L. Martin. Edited by Ambrose Stevens. Illustrated with engravings drawn from life.

Youatt on Sheep; 75

Their Breed, Management and Diseases, with illustrative engravings; to which are added Remarks on the Breeds and Management of Sheep in the United States, and on the Culture of Fine Wool in Silesia. By William Youatt.

American Architect. 6 00

The American Architect, comprising Original Designs of cheap Country and Village Residences, with Details, Specifications, Plans, and Directions, and an estimate of the Cost of each Design. By John W. Ritch, Architect. First and Second Series, quarto, bound in 1 vol., half roan.

Domestic Medicine. 3 00

Gunn's Domestic Medicine, or Poor Man's Friend, in the Hours of Affliction, Pain, and Sickness, Raymond's new revised edition, improved and enlarged. By John C. Gunn, 8vo.

Pedder's (James) Farmer's Land Measurer; . . . 50

Or, Pocket Companion; showing at one view, the Contents of any Piece of Land from Dimensions taken in Yards. With a set of Useful Agricultural Tables.

Chemical Field Lectures for Agriculturists; . . . 1 00

Or, Chemistry without a Master. By Dr. Julius Adolphus Stockhardt, Professor in the Royal Academy of Agriculture at Tharant. Translated from the German. Edited, with notes, by James E. Teschemacher.

Thaer's (Albert D.) Agriculture. 2 00

The Principles of Agriculture, by Albert D. Thaer; translated by William Shaw and Cuthbert W. Johnson, Esq., F.R.S. With a Memoir of the Author. 1 vol. 8vo.

This work is regarded by those who are competent to judge, as one of the most beautiful works that has ever appeared on the subject of agriculture. At the same time that it is eminently practical, it is philosophical, and, even to the general reader, remarkably entertaining.

Von Thaer was educated for a physician; and, after reaching the summit of his profession, he retired into the country, where his garden soon became the admiration of the citizens; and when he began to lay out plantations and orchards, to cultivate herbage and vegetables, the whole country was astonished at his science in the art of cultivation. He soon entered upon a large farm, and opened a school for the study of Agriculture, where his fame became known from one end of Europe to the other.

This great work of Von Thaer s has passed through ten editions in the United States, but it is still comparatively unknown. The attention of owners of landed estates in cities and towns, as well as those persons engaged in the practical pursuits of agriculture, is earnestly requested to this volume.

Guenon on Milch Cows; $0 60

A Treatise on Milch Cows, whereby the Quality and Quantity of Milk which any Cow will give may be accurately determined by observing Natural Marks or External Indications alone; the length of time she will continue to give Milk, &c., &c. By M. Francis Guenon, of Libourne, France. Translated by Nicholas P. Trist, Esq.; with Introduction, Remarks and Observations on the Cow and the Dairy, by John S. Skinner. Illustrated with numerous Engravings. Neatly done up in paper covers, 37 cts.

American Poultry Yard; 1 00

Comprising the Origin, History, and Description of the different Breeds of Domestic Poultry, with complete directions for their Breeding, Crossing, Rearing, Fattening, and Preparation for Market; including specific directions for Caponizing Fowls, and for the Treatment of the Principal Diseases to which they are subject, drawn from authentic sources and personal observation. Illustrated with numerous engravings. By D. J. Browne.

The Shepherd's Own Book: 2 00

With an Account of the different Breeds and Management and Diseases of Sheep, and General Directions in regard to Summer and Winter Management, Breeding, and the Treatment of Diseases; with illustrative engravings, by Youatt & Randall, embracing Skinner's Notes on the Breed and Management of Sheep in the United States, and on the Culture of Fine Wool.

Allen's (L. F.) Rural Architecture; 1 25

Being a complete Description of Farm Houses, Cottages, and Out Buildings, comprising Wood-houses, Workshops, Tool-houses, Carriage and Wagon houses, Stables, Smoke and Ash houses, Ice houses, Apiaries or Bee-houses, Poultry-houses, Rabbitry, Dovecote, Piggery, Barns, and Sheds for Cattle, &c., &c.; together with Lawns, Pleasure Grounds, and Parks; the Flower, Fruit, and Vegetable Garden; also, Useful and Ornamental Domestic Animals for the Country Resident, &c., &c. Also, the best method of conducting water into Cattle Yards and Houses. Beautifully Illustrated.

Allen's (R. L.) American Farm Book. 1 00

The American Farm Book; or, a Compend of American Agriculture, being a Practical Treatise on Soils, Manures, Draining, Irrigation, Grasses, Grain, Roots, Fruits, Cotton, Tobacco, Sugar Cane, Rice, and every Staple Product of the United States; with the best methods of planting, cultivating, and preparation for market. Illustrated by more than 100 engravings. By R. L. Allen.

Reemelin's (Chas.) Vine-dresser's Manual 50

An Illustrated Treatise on Vineyards and Wine-making, containing full instructions as to location and soil; preparation of ground; selection and propagation of vines; the treatment of a young Vineyard; trimming and training the vines; manures; and the making of wine.

Bement's (C. N.) Rabbit Fancier; 50

A Treatise on the Breeding, Rearing, Feeding and General Management of Rabbits, with remarks upon their diseases and remedies, to which are added full directions for the construction of Hutches, Rabbitries, &c., together with recipes for cooking and dressing for the table. Beautifully illustrated.

The Horse's Foot, and how to keep it Sound; . . 50

With Cuts Illustrating the Anatomy of the Foot, and containing valuable hints on shoeing and stable management in health and in disease. By William Miles.

Stephens' (Henry) Book of the Farm; **4 00**

A Complete Guide to the Farmer, Steward, Ploughman, Cattleman, Shepherd, Field Worker, and Dairy Maid. By Henry Stephens. With Four Hundred and Fifty Illustrations; to which are added Explanatory Notes, Remarks, &c., by J. S. Skinner. Really one of the best books for a farmer to possess.

Allen's (R. L.) Diseases of Domestic Animals; **75**

Being a History and Description of the Horse, Mule, Cattle, Sheep, Swine, Poultry, and Farm Dogs, with Directions for their Management, Breeding, Crossing, Rearing, Feeding, and Preparation for a profitable Market; also, their Diseases and Remedies, together with full Directions for the Management of the Dairy, and the Comparative Economy and Advantages of Working Animals, the Horse, Mule, Oxen, &c. By R. L. Allen.

Browne's (D. J.) American Bird Fancier; **50**

Considered with reference to the Breeding, Rearing, Feeding, Management, and Peculiarities of Cage and House Birds. Illustrated with Engravings. By D. Jay Browne.

Saxton's Rural Hand Books, ... **1 25**

First Series, containing Treatises on—

The Horse,
The Hog,
The Honey Bee,
The Pests of the Farm,
Domestic Fowls, and
The Cow.

Second Series, containing— **1 25**

Every Lady Her Own Flower Gardener,
Elements of Agriculture,
Bird Fancier,
Essay on Manures.
American Kitchen Gardener.
American Rose Culturist.

Third Series, containing— **1 25**

Miles on the Horse's Foot,
The Rabbit Fancier,
Weeks on Bees,
Vine-Dresser's Manual,
Bee-Keeper's Chart,
Chemistry made Easy.

Fourth Series, containing— **1 25**

Persoz on the Vine,
Liebig's Familiar Letters,
Hooper's Dog and Gun,
Skilful Housewife,
Browne's Memoirs of Indian Corn.

Boussingault's (J. B.) Rural Economy, ... **1 25**

In its Relations with Chemistry, Physics, and Meteorology; or, Chemistry applied to Agriculture. By J. B. Boussingault. Translated, with Notes, etc., by George Law, Agriculturist.

"The work is the fruit of a long life of study and experiment, and its perusal will aid the farmer greatly in obtaining a practical and scientific knowledge of his profession."

Thompson (R. D.) on the Food of Animals. **75**

Experimental Researches on the Food of Animals and the Fattening of Cattle; with remarks on the Food of Man. Based upon experiments undertaken by order of the British Government, by Robert Dundas Thompson, M.D., Lecturer on Practical Chemistry, University of Glasgow.

Richardson on Dogs: their Origin and Varieties. . **$0 50**

Directions as to their General Management. With numerous original anecdotes. Also, Complete Instructions as to Treatment under Disease. By H. D. Richardson. Illustrated with numerous wood engravings.

This is not only a cheap, but one of the best works ever published on the Dog.

Liebig's (Justus) Familiar Letters on Chemistry, . . **50**

And its relation to Commerce, Physiology, and Agriculture. Edited by John Gardener, M.D.

The Dog and Gun. **50**

A few Loose Chapters on Shooting, among which will be found some anecdotes and incidents. Also instructions for Dog Breaking, and interesting letters from Sportsmen. By A Bad Shot.

Morgan Horses; **1 25**

A Premium Essay on the Origin, History, and Characteristics of this remarkable American Breed of Horses; tracing the Pedigree from the original Justin Morgan, through the most noted of his Progeny, down to the present time. With numerous portraits. To which are added hints for Breeding, Breaking, and General Use and Management of Horses, with practical Directions for training them at Agricultural Fairs. By D. C. Linsley.

Bridgeman's (Thos.) Young Gardener's Assistant: . **1 50**

In three parts, Containing Catalogues of Garden and Flower Seed, with practical directions under each head for the Cultivation of Culinary Vegetables and Flowers. Also, directions for cultivating Fruit Trees, the Grape Vine, &c., to which is added a Calendar to each part, showing the work necessary to be done in departments each month of the year. One volume octavo.

Bridgeman's Kitchen Gardener's Instructor	½ Cloth	**50**
" " " "	Cloth	**60**
Bridgeman's Florist's Guide	½ Cloth	**50**
" " "	Cloth	**60**
Bridgeman's Fruit Cultivator's Manual	½ Cloth	**50**
" " " "	Cloth	**60**
Chinese Sugar Cane and Sugar Making; .	Cloth	**50**

Its History, Culture and Adaptation to the Soil, Climate and Economy of the United States, with an account of various processes of Manufacturing Sugar. Drawn from authentic sources by Charles F. Suansbury, A.M., late Commissioner at the Exhibition of all Nations at London.

The Cotton Planter's Manual; **1 00**

Being a Compilation of Facts from the Best Authorities on the Culture of Cotton, its Natural History, Chemical Analysis, Trade and Consumption; and embracing a History of Cotton and the Cotton Gin. By J. A. Turner.

Cobbett's American Gardener; **50**

A Treatise on the Situation, Soil, and Laying-Out of Gardens, and the Making and Managing of Hot Beds and Green-Houses, and on the Propagation and Cultivation of the several sorts of Vegetables, Herbs, Fruits and Flowers.

The Western Fruit Book; **1.25**

Being a Compend of the History, Modes of Propagation, Culture, &c., of Fruit Trees and Shrubs, &c., &c. By F. R. Elliott.

SAXTON'S

Hand Books of Rural and Domestic Economy.

All arranged and adapted to the Use of American Farmers.

PRICE 25 CENTS EACH.

Hogs;

THEIR ORIGIN AND VARIETIES; Management, With a View to Profit, and Treatment under Disease; also, Plain Directions relative to the most approved modes of preserving their Flesh. By H. D. RICHARDSON, author of "The Hive and the Honey Bee," &c., &c. With illustrations—12mo.

The Hive and the Honey Bee;

WITH PLAIN DIRECTIONS FOR OBTAINING A CONSIDERABLE ANNUAL Income from this branch of Rural Economy; also an Account of the Diseases of Bees and their Remedies, and Remarks as to their Enemies, and the best mode of protecting the Hives from their attacks. By H. D. RICHARDSON. With illustrations.

Domestic Fowls;

THEIR NATURAL HISTORY, BREEDING, REARING AND GENERAL Management. By H. D. RICHARDSON, author of "The Natural History of the Fossil Deer," &c. With illustrations.

The Horse;

THEIR ORIGIN AND VARIETIES; WITH PLAIN DIRECTIONS AS TO the Breeding, Rearing and General Management, with Instructions as to the Treatment of Disease. Handsomely Illustrated—12mo. By H. D. RICHARDSON.

The Rose;

THE AMERICAN ROSE CULTURIST; being a Practical Treatise on the Propagation, Cultivation, and Management in all Seasons, &c. With full directions for the Treatment of the Dahlia.

The Pests of the Farm;

WITH INSTRUCTIONS FOR THEIR EXTIRPATION; being a Manual of Plain Directions for the certain Destruction of every description of Vermin. With numerous illustrations on Wood.

An Essay on Manures;

SUBMITTED TO THE TRUSTEES OF THE MASSACHUSETTS SOCIETY for Promoting Agriculture, for their Premium. By SAMUEL H. DANA.

The American Bird Fancier;

CONSIDERED WITH REFERENCE TO THE BREEDING, REARING, FEEDing, Management and Peculiarities of Cage and House Birds. Illustrated with Engravings. By D. JAY BROWNE.

Chemistry Made Easy.

FOR THE USE OF FARMERS. By J. TOPHAM.

Elements of Agriculture.

TRANSLATED FROM THE FRENCH, and Adapted to the use of American Farmers. By F. G. SKINNER.

The American Kitchen Gardener;

Containing Directions for the Cultivation of Vegetables and Garden Fruits. By T. G. Fessenden.

Chinese Sugar-Cane and Sugar Making.

Its History, Culture and Adaptation to the Soil, Climate, and Economy of the United States, with an account of various processes of Manufacturing Sugar. Drawn from authentic sources by Charles F. Stansbury, A.M., late Commissioner at the Exhibition of all Nations at London.

Persoz' Culture of the Vine.

A New Process for the Culture of the Vine, by Persoz, Professor to the Faculty of Sciences of Strasbourg; directing Professor of the School of Pharmacy of the same city. Translated by J. O'C. Barclay, Surgeon U. S. N.

Browne's Memoirs of Indian Corn.

Joel Barlow's Poem on Hasty Pudding.

The Bee-keeper's Chart;

Being a brief Practical Treatise on the Instinct, Habits, and Management of the Honey Bee, in all its various Branches, the result of many years' practical experience, whereby the author has been enabled to divest the subject of much that has been considered mysterious and difficult to overcome, and render it more sure, profitable, and interesting to every one than it has heretofore been. By E. W. Phelps.

Every Lady Her Own Flower Gardener;

Addressed to the Industrious and Economical only; containing Simple and Practical Directions for Cultivating Plants and Flowers: also, Hints for the Management of Flowers in Rooms, with brief Botanical Descriptions of Plants and Flowers. The whole in plain and simple language. By Louisa Johnson.

The Cow; Dairy Husbandry and Cattle Breeding.

By M. M. Milburn, and revised by H. D. Richardson and Ambrose Stevens. With Illustrations.

Wilson on the Culture of Flax;

Its Treatment, Agricultural and Technical; delivered before the New York State Agricultural Society, at the Annual Fair at Saratoga, in September last, by John Wilson, late President of the Royal Agricultural College at Cirencester, England.

Weeks on Bees.—A Manual.

Or, an Easy Method of Managing Bees in the most profitable manner to their owner, with infallible rules to prevent their destruction by the Moth; with an Appendix by Wooster A. Flanders.

Reemelin's (Chas.) Vine-dresser's Manual;

Containing full Instructions as to Location and Soil; Preparation of Ground; Selection and Propagation of Vines; the Treatment of a Young Vineyard; trimming and training the vines; manures and the making of wine. Every department illustrated.

Bement's (C. N.) Rabbit Fancier.

A Treatise on the Breeding, Rearing, Feeding, and General Management of Rabbits, with remarks upon their diseases and remedies; to which are added full directions for the construction of Hutches, Rabbitries, &c., together with recipes for cooking and dressing for the table.

The Horse's Foot, and how to keep it Sound.

WITH CUTS ILLUSTRATING THE ANATOMY OF THE FOOT, and containing valuable hints on Shoeing and Stable Management both in Health and Disease. By WILLIAM MILES.

The Skillful Housewife.

OR, COMPLETE GUIDE TO DOMESTIC COOKERY, TASTE, COMFORT AND Economy, embracing 659 receipts pertaining to Household Duties, the care of Health, Gardening, Birds, Education of Children, etc., etc. By Mrs. L. G. ABELL.

Richardson on Dogs; their Origin and Varieties.

DIRECTIONS AS TO THEIR GENERAL MANAGEMENT. With numerous original anecdotes. Also, Complete Instructions as to Treatment under Disease. Illustrated with numerous wood engravings.

This is not only a cheap, but one of the best works ever published on the Dog.

Liebig's (Justus) Familiar Letters on Chemistry.

AND ITS RELATION TO COMMERCE, PHYSIOLOGY, AND AGRICULTURE. Edited by JOHN GARDENER, M.D.

The Dog and Gun.

A FEW LOOSE CHAPTERS ON SHOOTING, among which will be found some anecdotes and incidents. Also, Instructions for Dog Breaking, and interesting letters from Sportsmen. By A BAD SHOT.

Saxton's Hand-Books, Fifty Cent Series, in Paper.

BUIST'S KITCHEN GARDEN,
PARDEE ON THE STRAWBERRY,
CHORLTON'S GRAPE GROWERS' GUIDE,
THOMPSON'S FOOD OF ANIMALS,
ALLEN'S DOMESTIC ANIMALS,
FIELD'S PEAR CULTURE.

www.ingramcontent.com/pod-product-compliance
Lightning Source LLC
LaVergne TN
LVHW020219110826
845151LV00003B/762